The

STAR
Almanac

for land surveyors

FOR THE YEAR 2002

Prepared by
HM Nautical Almanac Office
on behalf of the Council for the Central
Laboratory of the Research Councils

LONDON: The Stationery Office

ISSN 0081–4377

ISBN 0 11 887317 2

CONTENTS

GREEK ALPHABET

α	alpha	ε	epsilon	ι	iota	ν	nu	ρ	rho	φ	phi
β	beta	ζ	zeta	κ	kappa	ξ	xi	σ	sigma	χ	chi
γ	gamma	η	eta	λ	lambda	ο	omicron	τ	tau	ψ	psi
δ	delta	θ	theta	μ	mu	π	pi	υ	upsilon	ω	omega

PREFACE, 2002

This fifty-second edition of the *Star Almanac* follows closely the model introduced for the year 1951, with some additions and changes introduced in the editions for 1973, 1977 and 1979.

This is the first edition that includes a CD-ROM which contains the *electronic version*. This e-book contains the almanac itself stored in Adobe's portable document format, and the coefficients that compactly represent the positions of the Sun and stars in ASCII files. Readers unable to use the CD-ROM and require the ASCII data files should contact this office.

The Star Almanac for Land Surveyors has been prepared by H.M. Nautical Almanac Office under the supervision of S.A. Bell; suggestions for additions and modifications should be sent to him at the address below.

Richard Holdaway
Director

Space Science and Technology Department,
Rutherford Appleton Laboratory,
Chilton, Didcot,
Oxfordshire, OX11 0QX, England

February 2001

THE STAR ALMANAC

The *Star Almanac for Land Surveyors* has been expressly designed to provide in a convenient form, and in a compact volume, the astronomical data required by surveyors. The time-argument is universal time (UT), and so the tabulated values may differ slightly from those in *The Astronomical Almanac*, in which the time-argument is dynamical time.

The main features of the Almanac are the tabulation of:

(a) *R*, the difference between Greenwich sidereal time (GHA Aries) and universal time;

(b) the declination of the Sun and *E*, the difference between the Greenwich hour angle of the Sun and universal time;

(c) the right ascensions and declinations of 695 stars, including all stars brighter than magnitude 4·0;

(d) extended refraction and other auxiliary tables.

The precision of tabulation varies between $1''$ and $0.'1$, and corresponds to the requirements of the land surveyor.

The following explanation is restricted to a brief outline of fundamental concepts and definitions, together with a description of the tabulated quantities and illustrations of how they are to be extracted from the Almanac. Reference must be made to the standard text-books for a description and illustrations of the method of use of the tabulated data.

TIME AND HOUR ANGLE

The position of a heavenly body in the sky, as seen by an observer on the Earth's surface, is most conveniently specified by its *altitude* above the horizon and its *azimuth* from the north. It is essentially a comparison between the observed and calculated values of altitude and azimuth that makes possible the determination of position and direction on the Earth's surface from astronomical observations. It is not practicable to tabulate the rapidly changing altitudes and azimuths of the heavenly bodies for all positions on the Earth, but they can be calculated for any position, by tables or by direct solution of a spherical triangle, from a knowledge of the geocentric position of the heavenly body referred to the Earth's equator.

In fact the quantities which the observer requires from the almanac are the *Greenwich hour angle* (GHA), namely the arc of the equator (or the angle at the pole) *westwards* from the Greenwich meridian, and the *declination* (Dec), or arc of the meridian from the equator to the body. Both are independent of the observer's position and can be tabulated directly. The GHA thus obtained will in practice either be combined with a known, or assumed, longitude to give *local hour angle* (LHA) or be used to derive longitude from a value of LHA deduced from observation.

The rotation of the Earth on its axis from *west* to *east* causes the heavenly bodies to *appear* to revolve round the Earth from *east* to *west*. Owing to the real motion of the Earth in its annual orbit about the Sun, the latter appears to alter its position with respect to the stars and so appears to revolve in a different period; moreover, owing to the varying motions of the Earth in its orbit, this period is not constant. Universal time (UT), which is the mean solar time of the meridian of Greenwich, is such that the average period of rotation of the Earth with respect to the Sun taken throughout the year is precisely 24^h; this is the *mean solar day*. (The mean solar day is now slightly longer than a day of 86 400 atomic seconds; the accumulated effect of this difference amounts to about 1 second each year.) The hour angle of the Sun increases at a rate which is, on the average, precisely that of mean solar time. The difference between the Greenwich hour angle of the Sun and universal time oscillates slowly and with small amplitude; it is convenient to tabulate this difference (denoted by E) and to obtain the Greenwich hour angle of the Sun from the relation:

$$\text{GHA Sun} = \text{UT} + E \tag{1}$$

in which multiples of 24^h may be omitted. The Sun transits over the Greenwich meridian when $\text{UT} = 24^h - E$.

The period of rotation of the Earth with respect to the stars is approximately $23^h\ 56^m\ 04\overset{s}{.}1$ of mean solar time, so that the hour angles of the stars increase more rapidly than that of the Sun. It is convenient to relate the positions of the stars to the point known as the vernal equinox, or the *first point of Aries* (abbreviated to "Aries"), which is the intersection of the Earth's equator and the ecliptic. The hour angle of this point increases by approximately $24^h\ 03^m\ 56\overset{s}{.}6$ each mean solar day. By analogy with the treatment of the Sun, a quantity R is defined such that it is the difference between GHA Aries and UT; R increases by $3^m\ 56\overset{s}{.}6$ a day. Thus

$$\text{GHA Aries} = \text{UT} + R \tag{2}$$

and is *Greenwich (apparent) sidereal time* (GST).

The hour angle of a star differs from that of Aries by the arc of the equator, or the angle at the pole, between the meridian of Aries and the meridian of the star; this difference is measured *eastwards* from the meridian of Aries to that of the star and is known as *right ascension* (RA).

The apparent right ascensions (and declinations) of stars are nearly constant, varying slightly owing to the real motions of the stars and to precession, and periodically owing to nutation and aberration. The GHA of a star is obtained from that of Aries by *subtracting* the right ascension:

$$\text{GHA Star} = \text{GHA Aries} - \text{RA Star} = \text{GST} - \text{RA Star} \tag{3}$$

whence the formula for the GHA of a star is

$$\text{GHA Star} = \text{UT} + R - \text{RA Star} \tag{4}$$

It will be seen that when a star (or other heavenly body) is on the Greenwich meridian, its right ascension is equal to the Greenwich sidereal time.

A knowledge of UT will enable the GHA of the Sun and stars to be calculated from tabulated values of E, R and RA of the stars. Standard radio time-signals give coordinated universal time (UTC), which differs from the international atomic-time scale by an exact number of seconds. Step adjustments are made as required (normally at 0^h on January 1 and July 1) so that the difference between the time-signal and UT does not exceed $0\overset{s}{.}9$. The user of this Almanac who requires to make and reduce observations to a precision of $0\overset{s}{.}1$ must therefore obtain the correction to the time-signal from coding in the signal or from other sources. Further information about time-signals is given on pages 60–61.

In the same way as UT, GHA and GST are related to the meridian of Greenwich, there exist mean solar times, hour angles and sidereal times related to the meridian of the observer and termed *local mean time* (LMT), *local hour angle* (LHA) and *local sidereal time* (LST). Each of these differs from the corresponding UT, GHA and GST by the longitude of the observer, if necessary expressed in time. Thus

$$\text{Longitude} = \text{LMT} - \text{UT} = \text{LHA} - \text{GHA} = \text{LST} - \text{GST} \tag{5}$$

where the sign convention east longitude is positive is used. Similarly,

$$\text{LMT} = \text{UT} + \text{longitude (east)}; \qquad \text{UT } = \text{LMT} - \text{longitude (east)} \tag{6}$$
$$\text{LHA } = \text{GHA} + \text{longitude (east)}; \qquad \text{LST} = \text{GST} + \text{longitude (east)} \tag{7}$$

Combination of (5) and (7) with (1) and (4), together with the tabulated declinations, will provide the data to enable positions to be determined from astronomical observations.

In practice, local mean and sidereal times are used primarily in connection with local phenomena such as rising and setting and meridian passage; as such they are mainly used for planning programmes. Just as the Greenwich meridian passage of the Sun occurs when $\text{UT} = 24^h - E$, the local meridian passage occurs when LMT is $24^h - E$; similarly a star transits over the local meridian when $\text{LST} = \text{RA}$. However, LHA Aries or LST, corresponding to the LHA of a star whose right ascension is zero, is frequently used as an argument for such tables as the Pole Star Table; it may be obtained by combining longitude with R and UT as:

$$\text{LST} = R + \text{UT} + \text{longitude (east)} \tag{8}$$

If mean time (UT or LMT) is required for a given sidereal time (GST or LST) it may be obtained from:

$$\text{UT} = \text{GST} - R; \quad \text{LMT} = \text{LST} - R \tag{9}$$

where R is to be taken for the (approximate) UT concerned, as explained at top of page xiii.

In *The Star Almanac* all hour angles are measured in time, so that longitudes expressed in arc must be converted to time, at the rate of 1^h for each $15°$, before being used in equations (5) to (8). 24^h may be added or subtracted as necessary in any of the equations (1) to (9); west longitudes may be treated as negative or subtracted from 24^h ($360°$) before use.

CRITICAL TABLES

Several of the tables in this volume are arranged as "critical" tables in which the tabulated values of the argument are those at which the respondent has a value midway between two tabular values; thus, each printed value of the respondent corresponds to the whole interval between two adjacent tabulated values of the argument. No interpolation is needed and the error cannot greatly exceed half the adopted interval, usually one unit. The "critical" tabular values are calculated in such a way that the upper of the two possible values of the respondent is to be taken.

For example, in the table for the interpolation of R on page 70, the increment of 16^s6 corresponds to any sidereal time interval between $1^h 40^m 44^s8$ and $1^h 41^m 21^s2$; if the sidereal time interval is the critical value $1^h 41^m 21^s2$ the upper (i.e. 16^s6) of the two possible values of the respondent is to be taken.

INTRODUCTION

DESCRIPTION OF CONTENTS

Ephemeris of the Sun (Pages 2–25)

On the upper part of these pages are tabulated for every 6^h of UT: the excess (R) of Greenwich sidereal time (GHA Aries) over universal time; the apparent declination of the Sun; and the excess (E) of GHA Sun over universal time. Since the mean difference of R is constant, provision for interpolation is made by a critical table on pages 70 to 72. The declination and E can be interpolated, by means of the table on the upper part of page 73 using the tabulated differences. These differences are printed in a column to the right of the function, on the half-line between consecutive rows, and are given in units of the last figure of the corresponding function, namely $0'.1$ for the declination and $0^s.1$ for E.

The average value of the Sun's semi-diameter is given for the half-month concerned, and should be used without interpolation.

The lower part of the page contains the local mean times of sunrise (left-hand page) or sunset (right-hand page) on every fifth day for latitudes between S 60° and N 60°. In order to permit tabulation for sufficient latitudes, the times are given to the nearest tenth of an hour only. This accuracy will be found adequate for most purposes.

The dates and times of the Moon's phases are given to assist in planning night observations.

Apparent Places of Stars (Pages 26–53)

Pages 26–51 contain the apparent right ascensions and declinations of 650 stars at monthly intervals throughout the year. They include all stars brighter than magnitude 4·0, and such other stars brighter than magnitude 4·5 as are tabulated in *Apparent Places of Fundamental Stars*, except close circumpolar stars; a few stars of magnitudes between 4·0 and 4·5 have been omitted, and the remainder included, to make the number exactly 650. Stars considered bright enough to be observed in daylight are indicated by the letter d following the star number on the left-hand page.

Pages 52–53 contain the apparent right ascensions and declinations of 35 supplementary fainter stars of declinations between 74° and 85°, together with 15 stars in this range already given on pages 26–51. Northern stars are given on the upper half of the page and southern stars on the lower half. No stars fainter than 5·5 have been included.

The positions given — to $0^s.1$ in right ascension and $1''$ in declination — are those for times near the beginning of the month, so chosen as to reduce to a minimum the error caused by using, for interpolation purposes, a fixed (nominal) month. An interpolation table is given on the lower part of page 73. It is to be entered with the day of the month as the vertical argument, and with the difference between the values for the month of observation and the following month as the horizontal argument.

Even though second differences are in some places large enough to be significant, their neglect will never lead to serious error; linear interpolation, as described, should invariably be used.

The stars are arranged in order of their right ascensions for 1975·0 and it is intended to keep this order unchanged for many years in spite of precessional changes. An index giving the number of a star whose name is known is given on pages 78–80.

The magnitudes and distances of fainter companions in double or multiple systems are given as footnotes on these pages.

Apparent Places of Circumpolar Stars (Pages 54–55)

On these pages are given, at 0^h UT for every ten days, the apparent positions of five northern and five southern circumpolar stars and also the Greenwich sidereal time.

The time of upper transit is found as the time (UT) when

$$GST = RA - \text{longitude (east)}$$

Since the GST is only tabulated to the nearest minute this equation may be used if the time is only required to the nearest minute. For greater accuracy use

$$UT = RA - R - \text{longitude (east)}$$

Polaris Table (Pages 56–59)

In this table are combined the data for obtaining latitude from an observed altitude of *Polaris* (suitably corrected for instrumental errors and refraction) and the azimuth of this star (measured from north, positive to the east and negative to the west), for all hour angles and northern latitudes. The six tabulated quantities, each given to a precision of $0'.1$, are a_0, a_1, a_2, referring to the correction to altitude, and b_0, b_1, b_2 to the azimuth.

$$\text{latitude} = \text{corrected observed altitude} + a_0 + a_1 + a_2$$
$$\text{azimuth} = (b_0 + b_1 + b_2) \sec (\text{latitude})$$

These quantities are based on the following formulae;

$$a_0 + a_1 + a_2 = -p \cos t + \frac{1}{2} p \sin p \sin^2 t \tan (\text{latitude})$$
$$b_0 + b_1 + b_2 = -p \sin t - p \sin p \sin t \cos t \tan (\text{latitude})$$

where

$$p = \text{polar distance of } Polaris = 90° - \text{declination}$$
$$t = \text{local hour angle of } Polaris = \text{LST} - \text{RA}$$

a_0, b_0, which depend upon LST only, include both terms of the above expressions for mean values of the position of *Polaris* and for a mean latitude of 50°. a_1, b_1 cater for the changes in the values of the second terms for latitudes differing from 50°, and are dependent on both LST and latitude. a_2, b_2 cater for the effects in the first terms of changes in the position of *Polaris* from its adopted mean, and are dependent on both LST and date.

The arrangement of the table is designed to avoid the necessity for entering, with the same LST, three separate tables, one of single entry and two of double entry. a_0, b_0 are tabulated for every 3^m of local sidereal time, and the remaining quantities for the mid-point of each hour of LST. a_1, b_1 are given for a range of latitude, and a_2, b_2 for each month during the year. In the latter case mean values are adopted for each month and no interpolation is necessary.

The table is to be entered with the LST of observation, which gives the values of a_0, b_0 directly; interpolation, with maximum differences of 8, can be done mentally. In the same vertical column, the values a_1, b_1 are found with the latitude, and those of a_2, b_2 with the date, as argument. Thus all six quantities can, if desired, be extracted together. The errors due to the adoption of a mean value of the LST for each of the subsidiary tables have been reduced to a minimum, and the total error is not likely to exceed $0'.2$. Interpolation between columns should not be attempted.

The observed altitude must be corrected for refraction (see pages 62–64) before being used for the determination of latitude. A table of natural secants is given on page 65 primarily to facilitate the calculation of the azimuth, which is measured from the north, positive to the east and negative to the west.

INTRODUCTION

Radio Time Signals (Pages 60–61)

These pages contain a list of the principal radio time signals used by surveyors.

Refraction Tables (Pages 62–64)

These tables give the correction for refraction in the form:

$$r = r_0 \times f$$

where r_0 is the mean refraction, taken with argument observed (strictly, apparent) altitude from the table on page 62, and f is a correcting factor to be found from the tables on pages 63 and 64 with arguments temperature and pressure. The tables of f are arranged so that for each tabulated value of pressure a critical table gives the corresponding factor with argument temperature. Interpolation of the tabulated temperatures to intermediate values of pressure may easily be performed, but is generally unnecessary.

The pressure is given in millibars; for pressures less than 970 millibars an equivalent height in metres is also given.

These tables are based on those of Harzer in *Publikation der Sternwarte in Kiel*, XIV, 1924; the values of mean refraction apply to a standard pressure of 754 mm and a standard temperature of 7° C. Within the range of altitude for which tabulations are given, there is no significant difference between the various theories that are extant.

Natural Secants (Page 65)

This table is intended mainly for use in the determination of azimuth from an observation of *Polaris* (see example on page xvi), for which purpose full three-decimal accuracy is not required. For angles less than 30° the table is given in critical form; above 30° values are tabulated for every 5′ and interpolation offers no difficulty.

Conversion to Degrees and to Days (Pages 66–67)

The first table enables minutes and seconds of arc to be converted to decimals of a degree; the second table facilitates conversion of hours, minutes and seconds into decimals of a day.

Table for Circum-Meridian Observations (Page 68)

This table gives the factor m for use in the reduction to the meridian of circum-meridian observations. The full correction is $A\,m$, where $A = \cos\phi\cos\delta\,\mathrm{cosec}\,\zeta$ and ϕ, δ, ζ are approximate values of latitude, declination, and zenith distance respectively. The factor m should be determined from this table for the time of each observation and, for each star, the mean of the values of m so found should be multiplied by the expression A above. Alternatively the expression $m = 0.0005454\,t^2$, where t is the hour angle in seconds of time, may be used to calculate m with a pocket calculator. The product $A\,m$ is then to be applied to the mean of the observed altitudes.

It should be noted that, particularly in low latitudes, the use of stars within 10° of the zenith is to be avoided.

Conversion of Time to Arc (Page 69)

This table needs no explanation; it can also easily be used for conversion of arc to time. An example of the use is given on page xiv.

Interpolation Table for R (Pages 70–72)

The principal purpose of this table is to provide for the interpolation of the quantity R, which is tabulated for every 6^h of UT in the ephemeris of the Sun (pages 2–25). The table gives, in critical form with respect to the left-hand argument (in bold type), the increment to be added to a tabular value of R for any given interval of UT. The interpolated value of R is added to the universal time to give the Greenwich sidereal time, as illustrated in the example on page xv.

This table also provides for the mutual conversion of intervals of mean solar and sidereal time. The increment corresponding to the left-hand argument is, in fact, the quantity to be added to that interval of mean solar time to convert it to the equivalent interval of mean sidereal time. The right-hand argument gives the decrement to be *subtracted* from that interval of mean sidereal time to convert it to the equivalent interval of mean solar time. (It is important to note that the right-hand argument does *not* give the sidereal-time interval corresponding to the left-hand solar-time argument, or vice-versa.)

Interpolation Table for Sun (Page 73)

This table serves for the interpolation of the values of the Sun's declination and *E* given on pages 2–25. The nearest value to the printed difference may usually be taken as the horizontal argument, and the time interval as the vertical argument. Interpolation for intermediate values of either argument may easily be performed, if required.

Interpolation Table for Stars (Page 73)

This table is to be used for interpolation of the apparent places of stars on pages 26–53. The day of the month on which the observation is made is to be used as the vertical argument, and the difference to the following month as the horizontal argument. The variation in length of the month is to be ignored, as mentioned on page x.

Polynomial Coefficients for R and Sun (Pages 74–75)

This table, for use with small programmable electronic calculators, provides polynomial coefficients of *R*, Dec, *E* and SD for the Sun, together with appropriate formulae to evaluate the data at any instant of time to an accuracy of $1''$ of arc. Explanatory notes and numerical illustrations are given on the left-hand page.

Star Charts (Pages 76–77)

The charts show the relative positions, as seen from the Earth, of all the stars tabulated in this Almanac. Stars of magnitude 3·5 or brighter, and some fainter stars, are identified by their Bayer letter. The numbers of such stars can be obtained directly from the index on pages 78–80. For other stars it is necessary to estimate the RA and Dec from the chart and then search the main tabulations to find the corresponding number. A fixed magnitude has been adopted for each variable star on the charts.

Index to Places of Stars (Pages 78–80)

This index, arranged in alphabetical order of constellations (genitive case), affords a rapid means of finding the number and thus the apparent place of a star known only by its name. Twenty-three stars having proper names in general use are listed on page 78, but no attempt has been made to give a complete list of such names.

ILLUSTRATIONS

The following examples illustrate the extraction from the Almanac of the data required by a surveyor when making astronomical observations.

A surveyor, whose approximate position is

longitude E 33° 32',   latitude N 35° 11'

makes a series of observations on 2002 January 6. On this date the Moon (from page 2) is at last quarter and will not therefore interfere with evening observations.

Sunrise, sunset. From pages 2 and 3, interpolating for latitude and date, the local mean times of sunrise and sunset are found to be 7^h2 and 17^h0 respectively. The approximate duration of nautical twilight in this latitude in winter is about an hour so that stellar observations are not likely to be possible after 6^h or before 18^h. Using the formula

$$LST = LMT + R$$

and taking out, from page 2, an approximate value of $R\ (= 7^h1)$, it is seen that the earliest LST of evening observation is

$$LST = 18^h0 + 7^h1 = 25^h1\ (= 1^h1)$$

An examination of the right ascensions and declinations of the stars on pages 26–55 will indicate which are the most suitable for observation on the meridian during the evening.

Sun. The Sun transits the local meridian when $LMT = 24^h - E$, i.e. $12^h\ 05^m\ 48^s$ approximately, where E is taken out for 12^h UT on January 6. To find the corresponding UT, the longitude must be converted to time and applied to LMT; using the time-arc conversion table on page 69,

$$33°\ 30' = 2^h\ 14^m; \quad 02' = 08^s; \quad \text{thus}\ 33°\ 32' = 2^h\ 14^m\ 08^s$$

whence, since the longitude is east,

$$UT = 12^h\ 05^m\ 48^s - 2^h\ 14^m\ 08^s = 9^h\ 51^m\ 40^s$$

It should be noted that this time is in error (in the present case by 2^s), as the value of E used should have been taken out for the UT of meridian passage; it is rarely necessary to apply any correction since the time is required only approximately.

Suppose that a meridian altitude of the Sun's lower limb is observed as $32°\ 04'6$ at UT $9^h\ 56^m\ 13^s$, the temperature being $6°$ C and the pressure 969 millibars. From page 2, the Sun's semi-diameter is found to be $16'3$ and the declination is obtained by interpolation as:

		d	h	m		°	'
page 2; 2002	January	6	06		S 22	30·7
page 73;			3	56		−1·2
					Sun's declination	S 22	29·5

The correction to the value for 6^h is obtained by entering the interpolation table on page 73 with $3^h\ 56^m$, the excess of UT over the previous tabular value, and 18, the tabular difference; the sign is obtained by inspection.

The mean refraction r_0 is seen from page 62 to be $93''$, while the correcting factor f is found from page 63 to be 0·97: the refraction correction, to be subtracted from the observed altitude, is thus $93'' \times 0·97 = 90'' = 1'5$. Strictly, a correction should be applied for the Sun's parallax, which may, for this purpose, be regarded as a constant; it suffices to add $0'1$ to all altitudes less than $70°$.

Thus the corrected observed altitude of the Sun's centre is

$$32°\ 04'6 + 16'3 - 1'5 + 0'1 = 32°\ 19'5$$

whence the deduced latitude is

$$90° - 22°\ 29'5 - 32°\ 19'5 = 35°\ 11'0$$

Later in the day an observation of the Sun is made at UT January $6^d\ 13^h\ 23^m\ 49^s3$; its GHA and Dec are found from page 2 as follows:

	d	h	m		h	m	s		°	'
2002 January 6				UT	13	23	49·3		°	'
page 2;		12		E	11	54	11·9	Dec	S 22	28·9
page 73;	1	24		diff. 66			−1·5	diff. 18		−0·4
				GHA	1	17	59·7	Dec	S 22	28·5

The signs of the corrections from page 73 have been put in by inspection; 24^h has been subtracted from the GHA.

Stars. For transit observations the LST of transit is equal to the right ascension; the corresponding approximate mean times of transit may be obtained as follows. The earliest UT for evening observations on 2002 January 6, is about 16^h; to a UT of 16^h there corresponds an LST of 1^h 18^m 21^s calculated as follows:

$$\text{On 2002 January } 6^d\ 16^h\ R = 7^h\ 03^m\ 34^s\ \text{(page 2)} + 39^s\ \text{(page 71)} = 7^h\ 04^m\ 13^s$$
$$GST = R + UT\qquad\quad = 23^h\ 04^m\ 13^s$$
$$LST = GST + \text{east long}\quad = 23^h\ 04^m\ 13^s + 2^h\ 14^m\ 08^s = 1^h\ 18^m\ 21^s$$

For a particular star, say α Ceti (No. 66), the RA is found from page 28, by simple interpolation, to be 3^h 02^m $22^s\!\cdot\!7$; transit therefore occurs later than 16^h UT by a sidereal time interval of 1^h 44^m 02^s; reference to the table on page 70 shows that the equivalent solar time interval is 17^s less than this, so that transit occurs at 17^h 43^m 45^s UT.

Alternatively the UT of transit may be found as follows:

		h	m	s
GST of transit = RA − east long = 3^h 02^m 23^s − 2^h 14^m 08^s	=	00	48	15
GST on 2002 January 6^d 12^h UT = $R + $ UT	=	19	03	34
Hence GST of transit is later than 12^h UT by an ST interval of		5	44	41
The decrement (page 72 using the right-hand argument) is				57
Hence UT of transit = 12^h + 5^h 44^m 41^s − 57^s	=	17	43	44

From page 29 the declination N $4°$ $05'$ $46''$ may be taken out without reference to the interpolation table on page 73.

The observed meridian altitude of α Ceti is $58°$ $55'$ $17''$, and the observed temperature and pressure are $8°$ C and 957 millibars, respectively. The mean refraction r_0 corresponding to the observed altitude is $35''$; entering the column on page 63 corresponding to the observed pressure of 957 millibars, it is seen that the observed temperature of $8°$ C is a critical value, so that the upper of the two possible values of f, namely 0·95, is taken. The refraction correction is thus $35'' \times 0·95 = 33''$ and the deduced latitude is

$$90° + 4°\ 05'\ 46'' - 58°\ 54'\ 44'' = 35°\ 11'\ 02''.$$

Later in the evening of 2002 January 6 several stars are observed out of the meridian; two of these are No. 2, β Cassiopeiæ, observed at 17^h 56^m $19^s\!\cdot\!1$ UT, and No. 212, *Procyon* (α CMi) observed at 18^h 53^m $26^s\!\cdot\!7$ UT.

Their GHA and Dec are obtained as:

		No. 2				No. 212		
		h	m	s		h	m	s
UT 2002 January 6^d		17	56	19·1		18	53	26·7
R (page 2)	12^h	7	03	34·3	18^h	7	04	33·4
Increment (pages 72, 70) 5^h 56^m $19^s\!\cdot\!1$				58·5	0^h 53^m $26^s\!\cdot\!7$			8·8
GST = GHA Aries		25	00	51·9		25	58	08·9
RA January (pages 26, 34)		0	09	15·8 ⎫		7	39	24·6 ⎫
Interpolation to January 6 (diff. −10)				−0·2 ⎭	(diff. +3)			+0·1 ⎭
GHA = GHA Aries − RA		0	51	36·3		18	18	44·2

		°	′	″		°	′	″
Dec, January (pages 27, 35)		N 59	09	49		N 5	13	12
Interpolation to January 6 (diff. −3)				−1	(diff. −3)			−1
Declination		N 59	09	48		N 5	13	11

In practice the increment to the RA and Dec will almost certainly be applied mentally, so that only the final values are written down.

Circumpolar stars. The time of upper transit of ζ Octantis on 2002 January 6, in longitude E 33° 32″ is found as follows:

	h	m
Approximate RA (page 55)	8	57
− Longitude (east)	−2	14
GST of upper transit = RA− longitude (east)	6	43
− GST on 2002 January 6ᵈ 00ʰ (page 55)	−7	02
ST interval from 0ʰ UT to upper transit (add 24ʰ because negative)	23	41
−ΔR, the correction* for an interval of 23ʰ 41ᵐ in ST (pages 70–72)		−4
UT of upper transit	23	37

* This correction is obtained using the right-hand argument of the table on pages 70–72 by combining three times the value of ΔR for 6ʰ with the value of ΔR for 5ʰ 41ᵐ.

For greater accuracy the value of R must be used as follows:

	h	m	s
RA (page 55)	8	56	41
−R on 2002 January 6ᵈ 00ʰ (page 2)	−7	01	36
− Longitude (east)	−2	14	08
ST interval (add 24ʰ because negative) = RA − R − longitude (east)	23	40	57
−ΔR, the correction for an interval of 23ʰ 40ᵐ 57ˢ in ST (pages 70–72)		−3	53
UT of upper transit	22	37	04

Pole Star. Earlier, observations had been made of *Polaris* for determination of latitude and azimuth; under the same conditions as the observation of α Ceti, the observed altitude was 35° 55′ 16″ at UT 17ʰ 46ᵐ 36ˢ.

The factor f is 0·95, the mean refraction (from page 62) is 80″ and the refraction correction is 76″.

	d	h	m	s		°	′		′
UT 2002 January	6	17	46	36	LST 3ʰ 05ᵐ 15ˢ	a_0 − 43·1	b_0 − 5·9
R 12ʰ		7	03	34	Latitude 35°	a_1 0·0	b_1 + 0·1
Increment 5ʰ 46ᵐ 36ˢ				57	January	a_2 + 0·1	b_2 0·0
GST		24	51	07	Sum	− 43·0	− 5·8
Longitude (east)		+2	14	08	Obs. Alt.	35 55·3	sec (lat)
					Refraction	− 1·3	= 1·223
LST		3	05	15			Azimuth
					Latitude	N 35 11·0	= −7·1

The calculation of LST requires no explanation; it may be remarked, however, that an accuracy of 1ˢ is ample. The Pole Star Table on page 56 is entered with the LST just obtained; the LST of 3ʰ 03ᵐ is the second entry in the fourth column on the top portion of the page and interpolation of a_0, b_0, for the 2ᵐ 15ˢ is done mentally. In the same column a_1 and b_1 are taken out, with simple interpolation, with argument latitude (here 35°); a_2 and b_2 are taken out directly with argument month (here January).

The sum of the corrections to altitude is added to the corrected observed altitude to give latitude. Similarly the contributions to the azimuth are summed and multiplied by sec (latitude) from page 65, to give the azimuth, which is positive to the east and negative to the west.

If greater accuracy is desired the RA and Dec of *Polaris* may be taken from page 54:

$$\text{RA} = 2^h\ 34^m\ 26^s \qquad \text{Dec} = \text{N}\ 89°\ 16′\ 36″$$

and the observation reduced by direct calculation.

MERCURY can be seen for a few days around greatest elongation, in the morning sky around February 21, June 21 and October 13 (the best conditions in northern latitudes occur in mid-October and in southern latitudes from mid-February to mid-March) and in the evening sky around January 11, May 4, September 1 and December 26 (the best conditions in northern latitudes occur from late April to early May and in southern latitudes from mid-August to mid-September).

VENUS is too close to the Sun for observation until late February when it appears as a brilliant object in the evening sky. During the last week of October it again becomes too close to the Sun for observation until the end of the first week of November when it reappears in the morning sky.

MARS can be seen in the evening sky in Aquarius at the beginning of the year. It moves into Pisces during the second week of January, into Aries in late February, Taurus in early April (passing 6° N of *Aldebaran* on April 29) and into Gemini from late May. It becomes too close to the Sun for observation during the second half of June, reappearing in the morning sky from late September in Leo. It then continues into Virgo in the first week of October (passing 3° N of *Spica* on November 20) and into Libra from mid-December.

JUPITER is at opposition on January 1 when it can be seen throughout the night in Gemini. Its eastward elongation then gradually decreases and from late March it can be seen only in the evening sky. In early July it becomes too close to the Sun for observation until early August when it reappears in the morning sky in Cancer. Its westward elongation gradually increases until by mid-November it can be seen for more than half the night passing into Leo in the second half of November and into Cancer from mid-December.

SATURN is in Taurus at the beginning of the year. It can be seen for more than half the night until late February after which it can be seen only in the evening sky. Its eastward elongation gradually decreases (passing 4° N of *Aldebaran* on March 31) and in the second half of May it becomes too close to the Sun for observation. It reappears in the morning sky in late June and passes into Orion at the beginning of September and Taurus in the second half of November. It is at opposition on December 17, when it can be seen throughout the night. For the remainder of the year its eastward elongation gradually decreases and is visible for the greater part of the night.

ECLIPSES, 2002

1. An annular eclipse of the Sun on June 10-11 is visible as a partial eclipse from eastern Asia, Japan, Indonesia, northern Australia, Pacific Ocean, northern Mexico, U.S.A. and Canada except the extreme north-east. The eclipse begins on June 10 at $20^h 52^m$ and ends on June 11 at $02^h 37^m$. The annular phase begins in the South China Sea on June 10 at $21^h 54^m$, crosses the central Pacific Ocean and ends on June 11 at $01^h 34^m$ off the western coast of Mexico. The maximum duration of the annular phase is $1^m 09^s$.

2. A total eclipse of the Sun on December 4. The path of totality begins in the South Atlantic Ocean passes through Angola, northern Botswana, southern Zimbabwe, Mozambique, Central Indian Ocean and ends in South Australia. The partial phase is visible from Africa except the north, S.E.Atlantic Ocean, central Indian Ocean, part of Antarctica, Indonesia, Australia and South Island of New Zealand. The eclipse begins at $04^h 51^m$ and ends at $10^h 11^m$; the total phase begins at $05^h 50^m$ and ends at $09^h 12^m$. The maximum duration of totality is $2^m 08^s$.

SUN – JANUARY, 2002

UT (d h)	R (h m s)	Dec (° ′)	E (h m s)	UT (d h)	R (h m s)	Dec (° ′)	E (h m s)
1 **0** Tues.	6 41 53·2	S23 02·0 $_{12}$	11 56 42·5 $_{70}$	**9** **0** Wed.	7 13 25·7	S22 09·4 $_{21}$	11 53 07·6 $_{63}$
6	42 52·4	23 00·8 $_{13}$	56 35·5 $_{71}$	**6**	14 24·8	22 07·3 $_{21}$	53 01·3 $_{62}$
12	43 51·5	22 59·5 $_{12}$	56 28·4 $_{70}$	**12**	15 24·0	22 05·2 $_{22}$	52 55·1 $_{62}$
18	44 50·7	22 58·3 $_{13}$	56 21·4 $_{70}$	**18**	16 23·1	22 03·0 $_{22}$	52 48·9 $_{62}$
2 **0** Wed.	6 45 49·8	S22 57·0 $_{14}$	11 56 14·4 $_{70}$	**10** **0** Thur.	7 17 22·2	S22 00·8 $_{22}$	11 52 42·7 $_{61}$
6	46 48·9	22 55·6 $_{13}$	56 07·4 $_{70}$	**6**	18 21·4	21 58·6 $_{22}$	52 36·6 $_{61}$
12	47 48·1	22 54·3 $_{14}$	56 00·4 $_{70}$	**12**	19 20·5	21 56·4 $_{23}$	52 30·5 $_{60}$
18	48 47·2	22 52·9 $_{14}$	55 53·4 $_{69}$	**18**	20 19·7	21 54·1 $_{23}$	52 24·5 $_{61}$
3 **0** Thur.	6 49 46·4	S22 51·5 $_{14}$	11 55 46·5 $_{69}$	**11** **0** Fri.	7 21 18·8	S21 51·8 $_{23}$	11 52 18·4 $_{60}$
6	50 45·5	22 50·1 $_{15}$	55 39·6 $_{69}$	**6**	22 17·9	21 49·5 $_{23}$	52 12·4 $_{59}$
12	51 44·6	22 48·6 $_{15}$	55 32·7 $_{69}$	**12**	23 17·1	21 47·2 $_{24}$	52 06·5 $_{59}$
18	52 43·8	22 47·1 $_{15}$	55 25·8 $_{68}$	**18**	24 16·2	21 44·8 $_{24}$	52 00·6 $_{59}$
4 **0** Fri.	6 53 42·9	S22 45·6 $_{15}$	11 55 19·0 $_{68}$	**12** **0** Sat.	7 25 15·4	S21 42·4 $_{24}$	11 51 54·7 $_{59}$
6	54 42·0	22 44·1 $_{16}$	55 12·2 $_{68}$	**6**	26 14·5	21 40·0 $_{25}$	51 48·8 $_{58}$
12	55 41·2	22 42·5 $_{16}$	55 05·4 $_{68}$	**12**	27 13·6	21 37·5 $_{24}$	51 43·0 $_{57}$
18	56 40·3	22 40·9 $_{16}$	54 58·6 $_{68}$	**18**	28 12·8	21 35·1 $_{25}$	51 37·3 $_{57}$
5 **0** Sat.	6 57 39·5	S22 39·3 $_{17}$	11 54 51·8 $_{67}$	**13** **0** Sun.	7 29 11·9	S21 32·6 $_{26}$	11 51 31·6 $_{57}$
6	58 38·6	22 37·6 $_{17}$	54 45·1 $_{67}$	**6**	30 11·1	21 30·0 $_{25}$	51 25·9 $_{57}$
12	6 59 37·7	22 35·9 $_{17}$	54 38·4 $_{67}$	**12**	31 10·2	21 27·5 $_{26}$	51 20·2 $_{57}$
18	7 00 36·9	22 34·2 $_{17}$	54 31·7 $_{66}$	**18**	32 09·4	21 24·9 $_{26}$	51 14·6 $_{56}$
6 **0** Sun.	7 01 36·0	S22 32·5 $_{18}$	11 54 25·1 $_{66}$	**14** **0** Mon.	7 33 08·5	S21 22·3 $_{26}$	11 51 09·0 $_{55}$
6	02 35·2	22 30·7 $_{18}$	54 18·5 $_{66}$	**6**	34 07·6	21 19·7 $_{27}$	51 03·5 $_{55}$
12	03 34·3	22 28·9 $_{18}$	54 11·9 $_{66}$	**12**	35 06·8	21 17·0 $_{27}$	50 58·0 $_{54}$
18	04 33·4	22 27·1 $_{19}$	54 05·3 $_{65}$	**18**	36 05·9	21 14·3 $_{27}$	50 52·6 $_{54}$
7 **0** Mon.	7 05 32·6	S22 25·2 $_{19}$	11 53 58·8 $_{65}$	**15** **0** Tues.	7 37 05·1	S21 11·6 $_{27}$	11 50 47·2 $_{54}$
6	06 31·7	22 23·3 $_{19}$	53 52·3 $_{65}$	**6**	38 04·2	21 08·9 $_{28}$	50 41·8 $_{53}$
12	07 30·8	22 21·4 $_{19}$	53 45·8 $_{64}$	**12**	39 03·3	21 06·1 $_{28}$	50 36·5 $_{53}$
18	08 30·0	22 19·5 $_{20}$	53 39·4 $_{65}$	**18**	40 02·5	21 03·3 $_{28}$	50 31·2 $_{52}$
8 **0** Tues.	7 09 29·1	S22 17·5 $_{20}$	11 53 32·9 $_{63}$	**16** **0** Wed.	7 41 01·6	S21 00·5 $_{28}$	11 50 26·0 $_{52}$
6	10 28·3	22 15·5 $_{20}$	53 26·6 $_{64}$	**6**	42 00·7	20 57·7 $_{28}$	50 20·8 $_{52}$
12	11 27·4	22 13·5 $_{20}$	53 20·2 $_{63}$	**12**	42 59·9	20 54·9 $_{29}$	50 15·6 $_{51}$
18	12 26·5	22 11·5 $_{21}$	53 13·9 $_{63}$	**18**	43 59·0	20 52·0 $_{29}$	50 10·5 $_{50}$
24	7 13 25·7	S22 09·4	11 53 07·6	**24**	7 44 58·2	S20 49·1	11 50 05·5

Sun's SD   16′·3

SUNRISE

Date	South Latitude								0°	North Latitude							
	60°	55°	50°	45°	40°	30°	20°	10°		10°	20°	30°	40°	45°	50°	55°	60°
Jan.	h	h	h	h	h	h	h	h	h	h	h	h	h	h	h	h	h
1	2·7	3·4	3·9	4·3	4·6	5·0	5·4	5·7	6·0	6·3	6·6	6·9	7·4	7·6	8·0	8·4	9·1
6	2·9	3·5	4·0	4·4	4·7	5·1	5·5	5·7	6·0	6·3	6·6	7·0	7·4	7·6	8·0	8·4	9·0
11	3·0	3·7	4·1	4·5	4·7	5·2	5·5	5·8	6·1	6·3	6·6	7·0	7·4	7·6	7·9	8·3	8·9
16	3·2	3·8	4·2	4·6	4·8	5·2	5·6	5·8	6·1	6·4	6·6	7·0	7·3	7·6	7·9	8·2	8·8
21	3·4	4·0	4·4	4·7	4·9	5·3	5·6	5·9	6·1	6·4	6·6	6·9	7·3	7·5	7·8	8·2	8·6
26	3·6	4·2	4·5	4·8	5·0	5·4	5·7	5·9	6·2	6·4	6·6	6·9	7·2	7·5	7·7	8·0	8·5
31	3·9	4·3	4·7	4·9	5·1	5·5	5·7	6·0	6·2	6·4	6·6	6·8	7·2	7·4	7·6	7·9	8·3

Moon's Phases: Last Quarter 6^{d} 03^{h} 55^{m} New Moon 13^{d} 13^{h} 29^{m}

UT		R	Dec	E	UT		R	Dec	E
d	h	h m s	° ′	h m s	d	h	h m s	° ′	h m s
17	0	7 44 58·2	S 20 49·1	11 50 05·5	**25**	0	8 16 30·6	S 19 03·4	11 47 48·3
Thur.	6	45 57·3	20 46·1 (30)	50 00·4 (51)	**Fri.**	6	17 29·7	18 59·8 (36)	47 44·8 (35)
	12	46 56·4	20 43·2 (29)	49 55·5 (49)		12	18 28·9	18 56·1 (37)	47 41·4 (34)
	18	47 55·6	20 40·2 (30)	49 50·5 (50)		18	19 28·0	18 52·3 (38)	47 38·0 (34)
			(30)	(48)				(37)	(33)
18	0	7 48 54·7	S 20 37·2	11 49 45·7	**26**	0	8 20 27·2	S 18 48·6	11 47 34·7
Fri.	6	49 53·9	20 34·2 (30)	49 40·8 (49)	**Sat.**	6	21 26·3	18 44·8 (38)	47 31·4 (33)
	12	50 53·0	20 31·1 (31)	49 36·0 (48)		12	22 25·4	18 41·0 (38)	47 28·2 (32)
	18	51 52·1	20 28·0 (31)	49 31·3 (47)		18	23 24·6	18 37·2 (38)	47 25·0 (32)
			(31)	(47)				(38)	(31)
19	0	7 52 51·3	S 20 24·9	11 49 26·6	**27**	0	8 24 23·7	S 18 33·4	11 47 21·9
Sat.	6	53 50·4	20 21·8 (31)	49 21·9 (47)	**Sun.**	6	25 22·9	18 29·6 (38)	47 18·9 (30)
	12	54 49·5	20 18·6 (32)	49 17·3 (46)		12	26 22·0	18 25·7 (39)	47 15·8 (31)
	18	55 48·7	20 15·5 (31)	49 12·8 (45)		18	27 21·1	18 21·8 (39)	47 12·9 (29)
			(32)	(45)				(39)	(29)
20	0	7 56 47·8	S 20 12·3	11 49 08·3	**28**	0	8 28 20·3	S 18 17·9	11 47 10·0
Sun.	6	57 47·0	20 09·0 (33)	49 03·8 (45)	**Mon.**	6	29 19·4	18 14·0 (39)	47 07·1 (29)
	12	58 46·1	20 05·8 (32)	48 59·4 (44)		12	30 18·6	18 10·0 (40)	47 04·3 (28)
	18	7 59 45·2	20 02·5 (33)	48 55·0 (44)		18	31 17·7	18 06·1 (39)	47 01·5 (28)
			(33)	(43)				(40)	(27)
21	0	8 00 44·4	S 19 59·2	11 48 50·7	**29**	0	8 32 16·8	S 18 02·1	11 46 58·8
Mon.	6	01 43·5	19 55·9 (33)	48 46·4 (43)	**Tues.**	6	33 16·0	17 58·0 (41)	46 56·1 (27)
	12	02 42·6	19 52·6 (33)	48 42·2 (42)		12	34 15·1	17 54·0 (40)	46 53·5 (26)
	18	03 41·8	19 49·2 (34)	48 38·0 (42)		18	35 14·3	17 50·0 (40)	46 51·0 (25)
			(34)	(41)				(41)	(25)
22	0	8 04 40·9	S 19 45·8	11 48 33·9	**30**	0	8 36 13·4	S 17 45·9	11 46 48·5
Tues.	6	05 40·1	19 42·4 (34)	48 29·8 (41)	**Wed.**	6	37 12·5	17 41·8 (41)	46 46·0 (25)
	12	06 39·2	19 39·0 (34)	48 25·8 (40)		12	38 11·7	17 37·7 (41)	46 43·6 (24)
	18	07 38·3	19 35·5 (35)	48 21·8 (40)		18	39 10·8	17 33·6 (41)	46 41·2 (24)
			(34)	(39)				(42)	(23)
23	0	8 08 37·5	S 19 32·1	11 48 17·9	**31**	0	8 40 10·0	S 17 29·4	11 46 38·9
Wed.	6	09 36·6	19 28·6 (35)	48 14·0 (39)	**Thur.**	6	41 09·1	17 25·2 (42)	46 36·7 (22)
	12	10 35·8	19 25·0 (36)	48 10·2 (38)		12	42 08·2	17 21·1 (41)	46 34·5 (22)
	18	11 34·9	19 21·5 (35)	48 06·4 (38)		18	43 07·4	17 16·9 (42)	46 32·3 (22)
			(36)	(37)				(43)	(21)
24	0	8 12 34·0	S 19 17·9	11 48 02·7	**32**	0	8 44 06·5	S 17 12·6	11 46 30·2
Thur.	6	13 33·2	19 14·3 (36)	47 59·0 (37)	**Fri.**	6	45 05·6	17 08·4 (42)	46 28·1 (21)
	12	14 32·3	19 10·7 (36)	47 55·4 (36)		12	46 04·8	17 04·1 (43)	46 26·1 (20)
	18	15 31·5	19 07·1 (36)	47 51·8 (36)		18	47 03·9	16 59·8 (43)	46 24·2 (19)
	24	8 16 30·6	S 19 03·4 (37)	11 47 48·3 (35)		24	8 48 03·1	S 16 55·5 (43)	11 46 22·3 (19)

Sun's SD   16′·3

SUNSET

Date	South Latitude								0°	North Latitude							
	60°	55°	50°	45°	40°	30°	20°	10°		10°	20°	30°	40°	45°	50°	55°	60°
Jan.	h	h	h	h	h	h	h	h	h	h	h	h	h	h	h	h	h
1	21·4	20·7	20·2	19·8	19·5	19·1	18·7	18·4	18·1	17·8	17·5	17·2	16·7	16·5	16·1	15·7	15·1
6	21·3	20·6	20·2	19·8	19·5	19·1	18·7	18·4	18·2	17·9	17·6	17·2	16·8	16·6	16·2	15·8	15·2
11	21·2	20·6	20·1	19·8	19·5	19·1	18·7	18·5	18·2	17·9	17·6	17·3	16·9	16·6	16·3	15·9	15·4
16	21·1	20·5	20·1	19·7	19·5	19·1	18·8	18·5	18·2	18·0	17·7	17·4	17·0	16·7	16·5	16·1	15·6
21	20·9	20·4	20·0	19·7	19·4	19·1	18·7	18·5	18·2	18·0	17·7	17·4	17·1	16·9	16·6	16·2	15·7
26	20·7	20·2	19·9	19·6	19·4	19·0	18·7	18·5	18·3	18·1	17·8	17·5	17·2	17·0	16·7	16·4	16·0
31	20·6	20·1	19·8	19·5	19·3	19·0	18·7	18·5	18·3	18·1	17·9	17·6	17·3	17·1	16·9	16·6	16·2

Moon's Phases: First Quarter 21^d 17^h 46^m Full Moon 28^d 22^h 50^m

SUN – FEBRUARY, 2002

UT		R	Dec	E	UT		R	Dec	E
d	h	h m s	° ′	h m s	d	h	h m s	° ′	h m s
1	0	8 44 06·5	S17 12·6 ₄₂	11 46 30·2 ₂₁	**9**	0	9 15 39·0	S14 47·9 ₄₈	11 45 49·0 ₅
Fri.	6	45 05·6	17 08·4 ₄₃	46 28·1 ₂₀	**Sat.**	6	16 38·1	14 43·1 ₄₈	45 48·5 ₄
	12	46 04·8	17 04·1 ₄₃	46 26·1 ₁₉		12	17 37·2	14 38·3 ₄₈	45 48·1 ₄
	18	47 03·9	16 59·8 ₄₃	46 24·2 ₁₉		18	18 36·4	14 33·5 ₄₉	45 47·7 ₄
2	0	8 48 03·1	S16 55·5 ₄₃	11 46 22·3 ₁₉	**10**	0	9 19 35·5	S14 28·6 ₄₈	11 45 47·3 ₃
Sat.	6	49 02·2	16 51·2 ₄₃	46 20·4 ₁₈	**Sun.**	6	20 34·7	14 23·8 ₄₉	45 47·0 ₂
	12	50 01·3	16 46·9 ₄₄	46 18·6 ₁₇		12	21 33·8	14 18·9 ₄₉	45 46·8 ₂
	18	51 00·5	16 42·5 ₄₄	46 16·9 ₁₇		18	22 32·9	14 14·0 ₄₉	45 46·6 ₁
3	0	8 51 59·6	S16 38·1 ₄₄	11 46 15·2 ₁₇	**11**	0	9 23 32·1	S14 09·1 ₄₉	11 45 46·5 ₁
Sun.	6	52 58·7	16 33·7 ₄₄	46 13·5 ₁₆	**Mon.**	6	24 31·2	14 04·2 ₅₀	45 46·4 ₁
	12	53 57·9	16 29·3 ₄₄	46 11·9 ₁₆		12	25 30·3	13 59·2 ₄₉	45 46·3 ₀
	18	54 57·0	16 24·9 ₄₄	46 10·3 ₁₅		18	26 29·5	13 54·3 ₅₀	45 46·3 ₁
4	0	8 55 56·2	S16 20·5 ₄₅	11 46 08·8 ₁₄	**12**	0	9 27 28·6	S13 49·3 ₅₀	11 45 46·4 ₁
Mon.	6	56 55·3	16 16·0 ₄₅	46 07·4 ₁₄	**Tues.**	6	28 27·8	13 44·3 ₄₉	45 46·5 ₁
	12	57 54·4	16 11·5 ₄₅	46 06·0 ₁₄		12	29 26·9	13 39·4 ₅₁	45 46·6 ₂
	18	58 53·6	16 07·0 ₄₅	46 04·6 ₁₃		18	30 26·0	13 34·3 ₅₀	45 46·8 ₃
5	0	8 59 52·7	S16 02·5 ₄₆	11 46 03·3 ₁₃	**13**	0	9 31 25·2	S13 29·3 ₅₀	11 45 47·1 ₂
Tues.	6	9 00 51·9	15 57·9 ₄₅	46 02·0 ₁₂	**Wed.**	6	32 24·3	13 24·3 ₅₁	45 47·3 ₄
	12	01 51·0	15 53·4 ₄₆	46 00·8 ₁₂		12	33 23·5	13 19·2 ₅₀	45 47·7 ₃
	18	02 50·1	15 48·8 ₄₆	45 59·6 ₁₁		18	34 22·6	13 14·2 ₅₁	45 48·0 ₅
6	0	9 03 49·3	S15 44·2 ₄₆	11 45 58·5 ₁₀	**14**	0	9 35 21·7	S13 09·1 ₅₁	11 45 48·5 ₄
Wed.	6	04 48·4	15 39·6 ₄₆	45 57·5 ₁₁	**Thur.**	6	36 20·9	13 04·0 ₅₁	45 48·9 ₆
	12	05 47·6	15 35·0 ₄₆	45 56·4 ₉		12	37 20·0	12 58·9 ₅₁	45 49·5 ₅
	18	06 46·7	15 30·4 ₄₇	45 55·5 ₉		18	38 19·1	12 53·8 ₅₁	45 50·0 ₆
7	0	9 07 45·8	S15 25·7 ₄₇	11 45 54·6 ₉	**15**	0	9 39 18·3	S12 48·7 ₅₂	11 45 50·6 ₇
Thur.	6	08 45·0	15 21·0 ₄₆	45 53·7 ₈	**Fri.**	6	40 17·4	12 43·5 ₅₁	45 51·3 ₇
	12	09 44·1	15 16·4 ₄₇	45 52·9 ₈		12	41 16·6	12 38·4 ₅₂	45 52·0 ₈
	18	10 43·3	15 11·7 ₄₈	45 52·1 ₇		18	42 15·7	12 33·2 ₅₂	45 52·8 ₇
8	0	9 11 42·4	S15 06·9 ₄₇	11 45 51·4 ₇	**16**	0	9 43 14·8	S12 28·0 ₅₁	11 45 53·5 ₉
Fri.	6	12 41·5	15 02·2 ₄₇	45 50·7 ₆	**Sat.**	6	44 14·0	12 22·9 ₅₂	45 54·4 ₉
	12	13 40·7	14 57·5 ₄₈	45 50·1 ₆		12	45 13·1	12 17·7 ₅₃	45 55·3 ₉
	18	14 39·8	14 52·7 ₄₈	45 49·5 ₅		18	46 12·2	12 12·4 ₅₂	45 56·2 ₁₀
	24	9 15 39·0	S14 47·9	11 45 49·0 ₅		24	9 47 11·4	S12 07·2	11 45 57·2

Sun's SD   16′2

SUNRISE

Date	South Latitude								0°	North Latitude							
	60°	55°	50°	45°	40°	30°	20°	10°		10°	20°	30°	40°	45°	50°	55°	60°
Feb.	h	h	h	h	h	h	h	h	h	h	h	h	h	h	h	h	h
1	3·9	4·4	4·7	4·9	5·2	5·5	5·7	6·0	6·2	6·4	6·6	6·8	7·2	7·3	7·6	7·9	8·2
6	4·2	4·5	4·8	5·1	5·2	5·5	5·8	6·0	6·2	6·4	6·6	6·8	7·1	7·2	7·5	7·7	8·1
11	4·4	4·7	5·0	5·2	5·3	5·6	5·8	6·0	6·2	6·3	6·5	6·7	7·0	7·1	7·3	7·5	7·8
16	4·6	4·9	5·1	5·3	5·5	5·7	5·9	6·0	6·2	6·3	6·5	6·7	6·9	7·0	7·2	7·3	7·6
21	4·8	5·1	5·3	5·4	5·5	5·7	5·9	6·0	6·2	6·3	6·4	6·6	6·8	6·9	7·0	7·2	7·4
26	5·0	5·3	5·4	5·5	5·7	5·8	5·9	6·0	6·2	6·3	6·4	6·5	6·7	6·7	6·8	7·0	7·1
31	5·3	5·4	5·5	5·7	5·7	5·9	6·0	6·1	6·2	6·2	6·3	6·4	6·5	6·6	6·7	6·8	6·9

Moon's Phases: Last Quarter 4ᵈ 13ʰ 33ᵐ   New Moon 12ᵈ 07ʰ 41ᵐ

UT		R	Dec	E	UT		R	Dec	E
d	h	h m s	° ′	h m s	d	h	h m s	° ′	h m s
17	**0**	9 47 11·4	S 12 07·2 (52)	11 45 57·2 (10)	**25**	**0**	10 18 43·8	S 9 14·2 (56)	11 46 51·0 (24)
Sun.	**6**	48 10·5	12 02·0 (53)	45 58·2 (11)	**Mon.**	**6**	19 43·0	9 08·6 (55)	46 53·4 (23)
	12	49 09·6	11 56·7 (52)	45 59·3 (11)		**12**	20 42·1	9 03·1 (56)	46 55·7 (25)
	18	50 08·8	11 51·5 (53)	46 00·4 (11)		**18**	21 41·2	8 57·5 (56)	46 58·2 (24)
18	**0**	9 51 07·9	S 11 46·2 (53)	11 46 01·5 (12)	**26**	**0**	10 22 40·4	S 8 51·9 (56)	11 47 00·6 (25)
Mon.	**6**	52 07·1	11 40·9 (53)	46 02·7 (13)	**Tues.**	**6**	23 39·5	8 46·3 (56)	47 03·1 (25)
	12	53 06·2	11 35·6 (53)	46 04·0 (13)		**12**	24 38·7	8 40·7 (56)	47 05·6 (26)
	18	54 05·3	11 30·3 (53)	46 05·3 (13)		**18**	25 37·8	8 35·1 (57)	47 08·2 (26)
19	**0**	9 55 04·5	S 11 25·0 (54)	11 46 06·6 (14)	**27**	**0**	10 26 36·9	S 8 29·4 (56)	11 47 10·8 (26)
Tues.	**6**	56 03·6	11 19·6 (53)	46 08·0 (14)	**Wed.**	**6**	27 36·1	8 23·8 (56)	47 13·4 (27)
	12	57 02·7	11 14·3 (54)	46 09·4 (14)		**12**	28 35·2	8 18·2 (57)	47 16·1 (27)
	18	58 01·9	11 08·9 (53)	46 10·8 (15)		**18**	29 34·3	8 12·5 (56)	47 18·8 (27)
20	**0**	9 59 01·0	S 11 03·6 (54)	11 46 12·3 (16)	**28**	**0**	10 30 33·5	S 8 06·9 (57)	11 47 21·5 (28)
Wed.	**6**	10 00 00·2	10 58·2 (54)	46 13·9 (16)	**Thur.**	**6**	31 32·6	8 01·2 (57)	47 24·3 (28)
	12	00 59·3	10 52·8 (54)	46 15·5 (16)		**12**	32 31·8	7 55·5 (56)	47 27·1 (28)
	18	01 58·4	10 47·4 (54)	46 17·1 (17)		**18**	33 30·9	7 49·9 (57)	47 29·9 (28)
21	**0**	10 02 57·6	S 10 42·0 (54)	11 46 18·8 (17)	**29**	**0**	10 34 30·0	S 7 44·2 (57)	11 47 32·7 (29)
Thur.	**6**	03 56·7	10 36·6 (54)	46 20·5 (18)	**Fri.**	**6**	35 29·2	7 38·5 (57)	47 35·6 (30)
	12	04 55·9	10 31·2 (55)	46 22·3 (17)		**12**	36 28·3	7 32·8 (57)	47 38·6 (29)
	18	05 55·0	10 25·7 (54)	46 24·0 (19)		**18**	37 27·4	7 27·1 (57)	47 41·5 (30)
22	**0**	10 06 54·1	S 10 20·3 (55)	11 46 25·9 (19)		**24**	10 38 26·6	S 7 21·4	11 47 44·5
Fri.	**6**	07 53·3	10 14·8 (54)	46 27·8 (19)					
	12	08 52·4	10 09·4 (55)	46 29·7 (19)					
	18	09 51·6	10 03·9 (55)	46 31·6 (20)					
23	**0**	10 10 50·7	S 9 58·4 (55)	11 46 33·6 (21)					
Sat.	**6**	11 49·8	9 52·9 (55)	46 35·7 (20)					
	12	12 49·0	9 47·4 (55)	46 37·7 (22)					
	18	13 48·1	9 41·9 (55)	46 39·9 (21)					
24	**0**	10 14 47·3	S 9 36·4 (56)	11 46 42·0 (22)					
Sun.	**6**	15 46·4	9 30·8 (55)	46 44·2 (22)					
	12	16 45·5	9 25·3 (55)	46 46·4 (23)					
	18	17 44·7	9 19·8 (56)	46 48·7 (23)					
	24	10 18 43·8	S 9 14·2	11 46 51·0					

Sun's SD   16ʹ2

SUNSET

Date	South Latitude								0°	North Latitude							
	60°	55°	50°	45°	40°	30°	20°	10°		10°	20°	30°	40°	45°	50°	55°	60°
Feb.	h	h	h	h	h	h	h	h	h	h	h	h	h	h	h	h	h
1	20·5	20·1	19·7	19·5	19·3	19·0	18·7	18·5	18·3	18·1	17·9	17·6	17·3	17·1	16·9	16·6	16·2
6	20·3	19·9	19·6	19·4	19·2	18·9	18·7	18·5	18·3	18·1	17·9	17·7	17·4	17·2	17·0	16·8	16·4
11	20·1	19·7	19·5	19·3	19·1	18·9	18·6	18·5	18·3	18·1	17·9	17·7	17·5	17·4	17·2	16·9	16·7
16	19·8	19·6	19·3	19·1	19·0	18·8	18·6	18·4	18·3	18·1	18·0	17·8	17·6	17·5	17·3	17·1	16·9
21	19·6	19·4	19·2	19·0	18·9	18·7	18·6	18·4	18·3	18·2	18·0	17·9	17·7	17·6	17·5	17·3	17·1
26	19·4	19·1	19·0	18·9	18·8	18·6	18·5	18·4	18·3	18·2	18·1	17·9	17·8	17·7	17·6	17·5	17·3
31	19·1	18·9	18·8	18·7	18·6	18·5	18·4	18·3	18·2	18·2	18·1	18·0	17·9	17·8	17·7	17·6	17·5

Moon's Phases: First Quarter 20ᵈ 12ʰ 02ᵐ   Full Moon 27ᵈ 09ʰ 17ᵐ

SUN – MARCH, 2002

UT		R	Dec	E	UT		R	Dec	E
d	h	h m s	° ′	h m s	d	h	h m s	° ′	h m s
1 Fri.	0	10 34 30·0	S 7 44·2 $_{57}$	11 47 32·7 $_{29}$	**9** Sat.	0	11 06 02·5	S 4 39·2 $_{59}$	11 49 19·1 $_{38}$
	6	35 29·2	7 38·5 $_{57}$	47 35·6 $_{30}$		6	07 01·6	4 33·3 $_{59}$	49 22·9 $_{37}$
	12	36 28·3	7 32·8 $_{57}$	47 38·6 $_{29}$		12	08 00·7	4 27·4 $_{58}$	49 26·6 $_{38}$
	18	37 27·4	7 27·1 $_{57}$	47 41·5 $_{30}$		18	08 59·9	4 21·6 $_{59}$	49 30·4 $_{38}$
2 Sat.	0	10 38 26·6	S 7 21·4 $_{57}$	11 47 44·5 $_{30}$	**10** Sun.	0	11 09 59·0	S 4 15·7 $_{59}$	11 49 34·2 $_{38}$
	6	39 25·7	7 15·7 $_{58}$	47 47·5 $_{31}$		6	10 58·2	4 09·8 $_{59}$	49 38·0 $_{39}$
	12	40 24·9	7 09·9 $_{57}$	47 50·6 $_{30}$		12	11 57·3	4 03·9 $_{58}$	49 41·9 $_{38}$
	18	41 24·0	7 04·2 $_{57}$	47 53·6 $_{31}$		18	12 56·4	3 58·1 $_{59}$	49 45·7 $_{39}$
3 Sun.	0	10 42 23·1	S 6 58·5 $_{58}$	11 47 56·7 $_{32}$	**11** Mon.	0	11 13 55·6	S 3 52·2 $_{59}$	11 49 49·6 $_{39}$
	6	43 22·3	6 52·7 $_{57}$	47 59·9 $_{31}$		6	14 54·7	3 46·3 $_{59}$	49 53·5 $_{39}$
	12	44 21·4	6 47·0 $_{58}$	48 03·0 $_{32}$		12	15 53·9	3 40·4 $_{59}$	49 57·4 $_{40}$
	18	45 20·5	6 41·2 $_{57}$	48 06·2 $_{32}$		18	16 53·0	3 34·5 $_{59}$	50 01·4 $_{39}$
4 Mon.	0	10 46 19·7	S 6 35·5 $_{58}$	11 48 09·4 $_{33}$	**12** Tues.	0	11 17 52·1	S 3 28·6 $_{59}$	11 50 05·3 $_{40}$
	6	47 18·8	6 29·7 $_{58}$	48 12·7 $_{32}$		6	18 51·3	3 22·7 $_{59}$	50 09·3 $_{40}$
	12	48 18·0	6 23·9 $_{57}$	48 15·9 $_{33}$		12	19 50·4	3 16·8 $_{59}$	50 13·3 $_{40}$
	18	49 17·1	6 18·2 $_{58}$	48 19·2 $_{34}$		18	20 49·5	3 10·9 $_{59}$	50 17·3 $_{40}$
5 Tues.	0	10 50 16·2	S 6 12·4 $_{58}$	11 48 22·6 $_{33}$	**13** Wed.	0	11 21 48·7	S 3 05·0 $_{59}$	11 50 21·3 $_{41}$
	6	51 15·4	6 06·6 $_{58}$	48 25·9 $_{34}$		6	22 47·8	2 59·1 $_{59}$	50 25·4 $_{40}$
	12	52 14·5	6 00·8 $_{58}$	48 29·3 $_{34}$		12	23 47·0	2 53·2 $_{60}$	50 29·4 $_{41}$
	18	53 13·7	5 55·0 $_{58}$	48 32·7 $_{34}$		18	24 46·1	2 47·2 $_{59}$	50 33·5 $_{41}$
6 Wed.	0	10 54 12·8	S 5 49·2 $_{58}$	11 48 36·1 $_{35}$	**14** Thur.	0	11 25 45·2	S 2 41·3 $_{59}$	11 50 37·6 $_{41}$
	6	55 11·9	5 43·4 $_{58}$	48 39·6 $_{35}$		6	26 44·4	2 35·4 $_{59}$	50 41·7 $_{41}$
	12	56 11·1	5 37·6 $_{59}$	48 43·1 $_{35}$		12	27 43·5	2 29·5 $_{59}$	50 45·8 $_{42}$
	18	57 10·2	5 31·7 $_{58}$	48 46·6 $_{35}$		18	28 42·6	2 23·6 $_{60}$	50 50·0 $_{41}$
7 Thur.	0	10 58 09·3	S 5 25·9 $_{58}$	11 48 50·1 $_{35}$	**15** Fri.	0	11 29 41·8	S 2 17·6 $_{59}$	11 50 54·1 $_{42}$
	6	10 59 08·5	5 20·1 $_{58}$	48 53·6 $_{36}$		6	30 40·9	2 11·7 $_{59}$	50 58·3 $_{42}$
	12	11 00 07·6	5 14·3 $_{59}$	48 57·2 $_{36}$		12	31 40·0	2 05·8 $_{59}$	51 02·5 $_{42}$
	18	01 06·8	5 08·4 $_{58}$	49 00·8 $_{36}$		18	32 39·2	1 59·9 $_{60}$	51 06·7 $_{42}$
8 Fri.	0	11 02 05·9	S 5 02·6 $_{59}$	11 49 04·4 $_{37}$	**16** Sat.	0	11 33 38·3	S 1 53·9 $_{59}$	11 51 10·9 $_{43}$
	6	03 05·0	4 56·7 $_{58}$	49 08·1 $_{36}$		6	34 37·5	1 48·0 $_{59}$	51 15·2 $_{42}$
	12	04 04·2	4 50·9 $_{59}$	49 11·7 $_{37}$		12	35 36·6	1 42·1 $_{59}$	51 19·4 $_{43}$
	18	05 03·3	4 45·0 $_{58}$	49 15·4 $_{37}$		18	36 35·7	1 36·2 $_{60}$	51 23·7 $_{43}$
	24	11 06 02·5	S 4 39·2	11 49 19·1		24	11 37 34·9	S 1 30·2	11 51 28·0 $_{43}$

Sun's SD   16′·1

SUNRISE

Date	South Latitude								0°	North Latitude							
	60°	55°	50°	45°	40°	30°	20°	10°		10°	20°	30°	40°	45°	50°	55°	60°
Mar.	h	h	h	h	h	h	h	h	h	h	h	h	h	h	h	h	h
1	5·2	5·4	5·5	5·6	5·7	5·8	6·0	6·1	6·2	6·2	6·3	6·5	6·6	6·7	6·7	6·8	7·0
6	5·4	5·5	5·6	5·7	5·8	5·9	6·0	6·1	6·1	6·2	6·3	6·3	6·5	6·5	6·6	6·7	6·7
11	5·6	5·7	5·8	5·8	5·9	6·0	6·0	6·1	6·1	6·2	6·2	6·2	6·3	6·3	6·4	6·4	6·5
16	5·8	5·9	5·9	6·0	6·0	6·0	6·0	6·1	6·1	6·1	6·1	6·2	6·2	6·2	6·2	6·2	6·2
21	6·0	6·0	6·0	6·0	6·0	6·1	6·1	6·1	6·1	6·1	6·1	6·0	6·0	6·0	6·0	6·0	6·0
26	6·2	6·2	6·2	6·2	6·1	6·1	6·1	6·1	6·0	6·0	6·0	6·0	5·9	5·9	5·8	5·8	5·7
31	6·4	6·4	6·3	6·3	6·2	6·2	6·1	6·1	6·0	6·0	5·9	5·8	5·8	5·7	5·7	5·6	5·5

Moon's Phases: Last Quarter 6^d 01^h 24^m New Moon 14^d 02^h 02^m

d	h	R (h m s)	Dec (° ′)	E (h m s)
17	0	11 37 34·9	S 1 30·2 (59)	11 51 28·0 (42)
Sun.	6	38 34·0	1 24·3 (59)	51 32·2 (43)
	12	39 33·1	1 18·4 (60)	51 36·5 (44)
	18	40 32·3	1 12·4 (59)	51 40·9 (43)
18	0	11 41 31·4	S 1 06·5 (59)	11 51 45·2 (43)
Mon.	6	42 30·6	1 00·6 (60)	51 49·5 (44)
	12	43 29·7	0 54·6 (59)	51 53·9 (43)
	18	44 28·8	0 48·7 (59)	51 58·2 (44)
19	0	11 45 28·0	S 0 42·8 (60)	11 52 02·6 (44)
Tues.	6	46 27·1	0 36·8 (59)	52 07·0 (44)
	12	47 26·2	0 30·9 (59)	52 11·4 (44)
	18	48 25·4	0 25·0 (60)	52 15·8 (44)
20	0	11 49 24·5	S 0 19·0 (59)	11 52 20·2 (44)
Wed.	6	50 23·7	0 13·1 (59)	52 24·6 (44)
	12	51 22·8	0 07·2 (59)	52 29·1 (45)
	18	52 21·9	S 0 01·3 (60)	52 33·5 (45)
21	0	11 53 21·1	N 0 04·7 (59)	11 52 38·0 (44)
Thur.	6	54 20·2	0 10·6 (59)	52 42·4 (45)
	12	55 19·4	0 16·5 (60)	52 46·9 (45)
	18	56 18·5	0 22·5 (59)	52 51·4 (44)
22	0	11 57 17·6	N 0 28·4 (59)	11 52 55·8 (45)
Fri.	6	58 16·8	0 34·3 (59)	53 00·3 (45)
	12	11 59 15·9	0 40·2 (59)	53 04·8 (45)
	18	12 00 15·1	0 46·1 (60)	53 09·3 (45)
23	0	12 01 14·2	N 0 52·1 (59)	11 53 13·8 (46)
Sat.	6	02 13·3	0 58·0 (59)	53 18·4 (45)
	12	03 12·5	1 03·9 (59)	53 22·9 (45)
	18	04 11·6	1 09·8 (59)	53 27·4 (45)
24	0	12 05 10·8	N 1 15·7 (59)	11 53 31·9 (46)
Sun.	6	06 09·9	1 21·6 (59)	53 36·5 (45)
	12	07 09·0	1 27·5 (59)	53 41·0 (46)
	18	08 08·2	1 33·4 (59)	53 45·6 (45)
	24	12 09 07·3	N 1 39·3	11 53 50·1
25	0	12 09 07·3	N 1 39·3 (59)	11 53 50·1 (46)
Mon.	6	10 06·4	1 45·2 (59)	53 54·7 (45)
	12	11 05·6	1 51·1 (59)	53 59·2 (46)
	18	12 04·7	1 57·0 (59)	54 03·8 (45)
26	0	12 13 03·9	N 2 02·9 (59)	11 54 08·3 (46)
Tues.	6	14 03·0	2 08·8 (59)	54 12·9 (46)
	12	15 02·1	2 14·7 (58)	54 17·5 (45)
	18	16 01·3	2 20·5 (59)	54 22·0 (46)
27	0	12 17 00·4	N 2 26·4 (59)	11 54 26·6 (46)
Wed.	6	17 59·6	2 32·3 (59)	54 31·2 (45)
	12	18 58·7	2 38·2 (58)	54 35·7 (46)
	18	19 57·8	2 44·0 (59)	54 40·3 (46)
28	0	12 20 57·0	N 2 49·9 (59)	11 54 44·9 (45)
Thur.	6	21 56·1	2 55·8 (58)	54 49·4 (46)
	12	22 55·2	3 01·6 (59)	54 54·0 (45)
	18	23 54·4	3 07·5 (58)	54 58·5 (46)
29	0	12 24 53·5	N 3 13·3 (59)	11 55 03·1 (46)
Fri.	6	25 52·6	3 19·2 (58)	55 07·7 (45)
	12	26 51·8	3 25·0 (58)	55 12·2 (45)
	18	27 50·9	3 30·8 (59)	55 16·7 (46)
30	0	12 28 50·1	N 3 36·7 (58)	11 55 21·3 (45)
Sat.	6	29 49·2	3 42·5 (58)	55 25·8 (46)
	12	30 48·3	3 48·3 (59)	55 30·4 (45)
	18	31 47·5	3 54·2 (58)	55 34·9 (45)
31	0	12 32 46·6	N 4 00·0 (58)	11 55 39·4 (45)
Sun.	6	33 45·7	4 05·8 (58)	55 43·9 (46)
	12	34 44·9	4 11·6 (58)	55 48·5 (45)
	18	35 44·0	4 17·4 (58)	55 53·0 (45)
32	0	12 36 43·2	N 4 23·2 (58)	11 55 57·5 (45)
Mon.	6	37 42·3	4 29·0 (58)	56 02·0 (44)
	12	38 41·4	4 34·8 (57)	56 06·4 (45)
	18	39 40·6	4 40·5 (58)	56 10·9 (45)
	24	12 40 39·7	N 4 46·3	11 56 15·4

Sun's SD   16ʹ·1

SUNSET

Date	South Latitude									North Latitude							
	60°	55°	50°	45°	40°	30°	20°	10°	0°	10°	20°	30°	40°	45°	50°	55°	60°
Mar.	h	h	h	h	h	h	h	h	h	h	h	h	h	h	h	h	h
1	19·2	19·0	18·9	18·8	18·7	18·6	18·4	18·4	18·3	18·2	18·1	18·0	17·9	17·8	17·7	17·6	17·4
6	18·9	18·8	18·7	18·6	18·6	18·5	18·4	18·3	18·2	18·2	18·1	18·0	17·9	17·9	17·8	17·7	17·6
11	18·7	18·6	18·6	18·5	18·4	18·4	18·3	18·3	18·2	18·2	18·1	18·1	18·0	18·0	18·0	17·9	17·9
16	18·4	18·4	18·4	18·3	18·3	18·3	18·2	18·2	18·2	18·2	18·2	18·1	18·1	18·1	18·1	18·1	18·1
21	18·2	18·2	18·2	18·2	18·2	18·2	18·2	18·2	18·2	18·2	18·2	18·2	18·2	18·2	18·2	18·2	18·3
26	17·9	18·0	18·0	18·0	18·0	18·1	18·1	18·1	18·1	18·2	18·2	18·2	18·3	18·3	18·4	18·4	18·5
31	17·7	17·8	17·8	17·9	17·9	18·0	18·0	18·1	18·1	18·2	18·2	18·3	18·4	18·4	18·5	18·6	18·7

Moon's Phases: First Quarter 22^d 02^h 28^m Full Moon 28^d 18^h 25^m

SUN – APRIL, 2002

UT	R	Dec	E	UT	R	Dec	E
d h	h m s	° '	h m s	d h	h m s	° '	h m s
1 0	12 36 43·2	N 4 23·2 (58)	11 55 57·5 (45)	**9** 0	13 08 15·6	N 7 25·6 (56)	11 58 16·0 (41)
Mon. 6	37 42·3	4 29·0 (58)	56 02·0 (44)	**Tues.** 6	09 14·7	7 31·2 (56)	58 20·1 (41)
12	38 41·4	4 34·8 (57)	56 06·4 (45)	12	10 13·9	7 36·8 (56)	58 24·2 (40)
18	39 40·6	4 40·5 (58)	56 10·9 (45)	18	11 13·0	7 42·4 (55)	58 28·2 (41)
2 0	12 40 39·7	N 4 46·3 (58)	11 56 15·4 (44)	**10** 0	13 12 12·2	N 7 47·9 (56)	11 58 32·3 (40)
Tues. 6	41 38·9	4 52·1 (58)	56 19·8 (45)	**Wed.** 6	13 11·3	7 53·5 (55)	58 36·3 (41)
12	42 38·0	4 57·9 (57)	56 24·3 (44)	12	14 10·4	7 59·0 (56)	58 40·4 (40)
18	43 37·1	5 03·6 (58)	56 28·7 (45)	18	15 09·6	8 04·6 (55)	58 44·4 (39)
3 0	12 44 36·3	N 5 09·4 (57)	11 56 33·2 (44)	**11** 0	13 16 08·7	N 8 10·1 (55)	11 58 48·3 (40)
Wed. 6	45 35·4	5 15·1 (58)	56 37·6 (44)	**Thur.** 6	17 07·8	8 15·6 (56)	58 52·3 (40)
12	46 34·6	5 20·9 (57)	56 42·0 (44)	12	18 07·0	8 21·2 (55)	58 56·3 (39)
18	47 33·7	5 26·6 (57)	56 46·4 (44)	18	19 06·1	8 26·7 (55)	59 00·2 (39)
4 0	12 48 32·8	N 5 32·3 (58)	11 56 50·8 (44)	**12** 0	13 20 05·2	N 8 32·2 (54)	11 59 04·1 (39)
Thur. 6	49 32·0	5 38·1 (57)	56 55·2 (44)	**Fri.** 6	21 04·4	8 37·6 (55)	59 08·0 (39)
12	50 31·1	5 43·8 (57)	56 59·5 (44)	12	22 03·5	8 43·1 (55)	59 11·9 (38)
18	51 30·3	5 49·5 (57)	57 03·9 (43)	18	23 02·7	8 48·6 (55)	59 15·7 (39)
5 0	12 52 29·4	N 5 55·2 (57)	11 57 08·2 (44)	**13** 0	13 24 01·8	N 8 54·1 (54)	11 59 19·6 (38)
Fri. 6	53 28·5	6 00·9 (57)	57 12·6 (43)	**Sat.** 6	25 00·9	8 59·5 (55)	59 23·4 (38)
12	54 27·7	6 06·6 (57)	57 16·9 (43)	12	26 00·1	9 05·0 (54)	59 27·2 (38)
18	55 26·8	6 12·3 (57)	57 21·2 (43)	18	26 59·2	9 10·4 (54)	59 31·0 (37)
6 0	12 56 26·0	N 6 18·0 (57)	11 57 25·5 (43)	**14** 0	13 27 58·3	N 9 15·8 (54)	11 59 34·7 (38)
Sat. 6	57 25·1	6 23·7 (56)	57 29·8 (43)	**Sun.** 6	28 57·5	9 21·2 (54)	59 38·5 (37)
12	58 24·2	6 29·3 (57)	57 34·1 (42)	12	29 56·6	9 26·6 (54)	59 42·2 (37)
18	12 59 23·4	6 35·0 (56)	57 38·3 (43)	18	30 55·8	9 32·0 (54)	59 45·9 (36)
7 0	13 00 22·5	N 6 40·6 (57)	11 57 42·6 (42)	**15** 0	13 31 54·9	N 9 37·4 (54)	11 59 49·5 (37)
Sun. 6	01 21·6	6 46·3 (56)	57 46·8 (42)	**Mon.** 6	32 54·0	9 42·8 (53)	59 53·2 (36)
12	02 20·8	6 51·9 (57)	57 51·0 (42)	12	33 53·2	9 48·1 (54)	11 59 56·8 (36)
18	03 19·9	6 57·6 (56)	57 55·2 (42)	18	34 52·3	9 53·5 (53)	12 00 00·4 (36)
8 0	13 04 19·1	N 7 03·2 (56)	11 57 59·4 (42)	**16** 0	13 35 51·5	N 9 58·8 (54)	12 00 04·0 (36)
Mon. 6	05 18·2	7 08·8 (56)	58 03·6 (41)	**Tues.** 6	36 50·6	10 04·2 (53)	00 07·6 (35)
12	06 17·3	7 14·4 (56)	58 07·7 (41)	12	37 49·7	10 09·5 (53)	00 11·1 (36)
18	07 16·5	7 20·0 (56)	58 11·8 (41)	18	38 48·9	10 14·8 (53)	00 14·7 (36)
24	13 08 15·6	N 7 25·6 (56)	11 58 16·0 (42)	24	13 39 48·0	N 10 20·1 (53)	12 00 18·1 (34)

Sun's SD 16·0

SUNRISE

Date	South Latitude											North Latitude					
	60°	55°	50°	45°	40°	30°	20°	10°	0°	10°	20°	30°	40°	45°	50°	55°	60°
Apr.	h	h	h	h	h	h	h	h	h	h	h	h	h	h	h	h	h
1	6·5	6·4	6·3	6·3	6·2	6·2	6·1	6·1	6·0	6·0	5·9	5·8	5·7	5·7	5·6	5·5	5·5
6	6·7	6·5	6·5	6·4	6·3	6·2	6·1	6·1	6·0	5·9	5·8	5·7	5·6	5·5	5·5	5·3	5·2
11	6·9	6·7	6·6	6·5	6·4	6·3	6·2	6·1	6·0	5·9	5·8	5·6	5·5	5·4	5·3	5·1	5·0
16	7·1	6·9	6·7	6·6	6·5	6·3	6·2	6·1	6·0	5·8	5·7	5·5	5·3	5·2	5·1	4·9	4·7
21	7·3	7·0	6·8	6·7	6·6	6·4	6·2	6·1	5·9	5·8	5·6	5·5	5·2	5·1	4·9	4·7	4·5
26	7·5	7·2	7·0	6·8	6·7	6·4	6·2	6·1	5·9	5·7	5·6	5·4	5·1	5·0	4·8	4·5	4·2
31	7·7	7·3	7·1	6·9	6·7	6·5	6·3	6·1	5·9	5·7	5·5	5·3	5·0	4·8	4·6	4·3	4·0

Moon's Phases: Last Quarter 4^d 15^h 29^m New Moon 12^d 19^h 21^m

UT d h	R (h m s)	Dec (° ′)	E (h m s)	UT d h	R (h m s)	Dec (° ′)	E (h m s)
17 0	13 39 48·0	N10 20·1 $_{53}$	12 00 18·1 $_{35}$	**25** 0	14 11 20·5	N13 03·6 $_{49}$	12 01 57·1 $_{26}$
Wed. 6	40 47·1	10 25·4 $_{53}$	00 21·6 $_{35}$	Thur. 6	12 19·6	13 08·5 $_{49}$	01 59·7 $_{26}$
12	41 46·3	10 30·7 $_{53}$	00 25·1 $_{34}$	12	13 18·7	13 13·4 $_{49}$	02 02·3 $_{26}$
18	42 45·4	10 36·0 $_{52}$	00 28·5 $_{34}$	18	14 17·9	13 18·3 $_{48}$	02 04·9 $_{26}$
18 0	13 43 44·6	N10 41·2 $_{53}$	12 00 31·9 $_{34}$	**26** 0	14 15 17·0	N13 23·1 $_{49}$	12 02 07·5 $_{25}$
Thur. 6	44 43·7	10 46·5 $_{52}$	00 35·3 $_{34}$	Fri. 6	16 16·1	13 28·0 $_{48}$	02 10·0 $_{25}$
12	45 42·8	10 51·7 $_{52}$	00 38·7 $_{33}$	12	17 15·3	13 32·8 $_{48}$	02 12·5 $_{24}$
18	46 42·0	10 56·9 $_{52}$	00 42·0 $_{33}$	18	18 14·4	13 37·6 $_{48}$	02 14·9 $_{25}$
19 0	13 47 41·1	N11 02·1 $_{52}$	12 00 45·3 $_{33}$	**27** 0	14 19 13·6	N13 42·4 $_{48}$	12 02 17·4 $_{24}$
Fri. 6	48 40·3	11 07·3 $_{52}$	00 48·6 $_{33}$	Sat. 6	20 12·7	13 47·2 $_{48}$	02 19·8 $_{24}$
12	49 39·4	11 12·5 $_{52}$	00 51·9 $_{32}$	12	21 11·8	13 52·0 $_{48}$	02 22·2 $_{23}$
18	50 38·6	11 17·7 $_{52}$	00 55·1 $_{32}$	18	22 11·0	13 56·8 $_{47}$	02 24·5 $_{23}$
20 0	13 51 37·7	N11 22·9 $_{51}$	12 00 58·3 $_{32}$	**28** 0	14 23 10·1	N14 01·5 $_{48}$	12 02 26·8 $_{23}$
Sat. 6	52 36·8	11 28·0 $_{52}$	01 01·5 $_{32}$	Sun. 6	24 09·2	14 06·3 $_{47}$	02 29·1 $_{22}$
12	53 36·0	11 33·2 $_{51}$	01 04·7 $_{31}$	12	25 08·4	14 11·0 $_{47}$	02 31·3 $_{22}$
18	54 35·1	11 38·3 $_{51}$	01 07·8 $_{31}$	18	26 07·5	14 15·7 $_{47}$	02 33·5 $_{22}$
21 0	13 55 34·2	N11 43·4 $_{51}$	12 01 10·9 $_{31}$	**29** 0	14 27 06·7	N14 20·4 $_{46}$	12 02 35·7 $_{21}$
Sun. 6	56 33·4	11 48·5 $_{51}$	01 14·0 $_{31}$	Mon. 6	28 05·8	14 25·0 $_{47}$	02 37·8 $_{22}$
12	57 32·5	11 53·6 $_{51}$	01 17·1 $_{30}$	12	29 04·9	14 29·7 $_{46}$	02 40·0 $_{20}$
18	58 31·7	11 58·7 $_{51}$	01 20·1 $_{30}$	18	30 04·1	14 34·3 $_{47}$	02 42·0 $_{21}$
22 0	13 59 30·8	N12 03·8 $_{50}$	12 01 23·1 $_{30}$	**30** 0	14 31 03·2	N14 39·0 $_{46}$	12 02 44·1 $_{20}$
Mon. 6	14 00 29·9	12 08·8 $_{51}$	01 26·1 $_{29}$	Tues. 6	32 02·4	14 43·6 $_{46}$	02 46·1 $_{20}$
12	01 29·1	12 13·9 $_{50}$	01 29·0 $_{30}$	12	33 01·5	14 48·2 $_{46}$	02 48·1 $_{19}$
18	02 28·2	12 18·9 $_{50}$	01 32·0 $_{29}$	18	34 00·6	14 52·8 $_{45}$	02 50·0 $_{19}$
23 0	14 03 27·4	N12 23·9 $_{50}$	12 01 34·9 $_{28}$	**31** 0	14 34 59·8	N14 57·3 $_{46}$	12 02 51·9 $_{19}$
Tues. 6	04 26·5	12 28·9 $_{50}$	01 37·7 $_{29}$	Wed. 6	35 58·9	15 01·9 $_{45}$	02 53·8 $_{18}$
12	05 25·6	12 33·9 $_{50}$	01 40·6 $_{28}$	12	36 58·1	15 06·4 $_{46}$	02 55·6 $_{18}$
18	06 24·8	12 38·9 $_{50}$	01 43·4 $_{28}$	18	37 57·2	15 11·0 $_{45}$	02 57·4 $_{18}$
24 0	14 07 23·9	N12 43·9 $_{49}$	12 01 46·2 $_{27}$	24	14 38 56·4	N15 15·5	12 02 59·2
Wed. 6	08 23·0	12 48·8 $_{50}$	01 48·9 $_{28}$				
12	09 22·2	12 53·8 $_{49}$	01 51·7 $_{27}$				
18	10 21·3	12 58·7 $_{49}$	01 54·4 $_{27}$				
24	14 11 20·5	N13 03·6 $_{49}$	12 01 57·1				

Sun's SD   15′·9

SUNSET

Date	South Latitude									North Latitude							
Apr.	60°	55°	50°	45°	40°	30°	20°	10°	0°	10°	20°	30°	40°	45°	50°	55°	60°
	h	h	h	h	h	h	h	h	h	h	h	h	h	h	h	h	h
1	17·6	17·7	17·8	17·8	17·9	17·9	18·0	18·1	18·1	18·2	18·2	18·3	18·4	18·4	18·5	18·6	18·7
6	17·4	17·5	17·6	17·7	17·7	17·9	17·9	18·0	18·1	18·2	18·3	18·4	18·5	18·6	18·6	18·8	18·9
11	17·1	17·3	17·4	17·5	17·6	17·7	17·9	18·0	18·1	18·2	18·3	18·4	18·6	18·7	18·8	18·9	19·1
16	16·9	17·1	17·3	17·4	17·5	17·7	17·8	17·9	18·1	18·2	18·3	18·5	18·6	18·8	18·9	19·1	19·3
21	16·7	16·9	17·1	17·2	17·4	17·6	17·7	17·9	18·0	18·2	18·3	18·5	18·7	18·9	19·0	19·2	19·5
26	16·4	16·7	16·9	17·1	17·3	17·5	17·7	17·9	18·0	18·2	18·4	18·6	18·8	19·0	19·2	19·4	19·7
31	16·2	16·6	16·8	17·0	17·1	17·4	17·6	17·8	18·0	18·2	18·4	18·6	18·9	19·1	19·3	19·6	19·9

Moon's Phases: First Quarter 20^d 12^h 48^m Full Moon 27^d 03^h 00^m

SUN – MAY, 2002

UT		R	Dec	E	UT		R	Dec	E
d	h	h m s	° ′	h m s	d	h	h m s	° ′	h m s
1	0	14 34 59·8	N14 57·3 (46)	12 02 51·9 (19)	9	0	15 06 32·2	N17 15·0 (40)	12 03 34·3 (8)
Wed.	6	35 58·9	15 01·9 (45)	02 53·8 (18)	Thur.	6	07 31·4	17 19·0 (40)	03 35·1 (6)
	12	36 58·1	15 06·4 (46)	02 55·6 (18)		12	08 30·5	17 23·0 (40)	03 35·7 (7)
	18	37 57·2	15 11·0 (45)	02 57·4 (18)		18	09 29·6	17 27·0 (40)	03 36·4 (6)
2	0	14 38 56·4	N15 15·5 (45)	12 02 59·2 (17)	10	0	15 10 28·8	N17 31·0 (39)	12 03 37·0 (6)
Thur.	6	39 55·5	15 20·0 (44)	03 00·9 (17)	Fri.	6	11 27·9	17 34·9 (39)	03 37·6 (6)
	12	40 54·6	15 24·4 (45)	03 02·6 (17)		12	12 27·1	17 38·8 (40)	03 38·2 (5)
	18	41 53·8	15 28·9 (44)	03 04·3 (16)		18	13 26·2	17 42·8 (38)	03 38·7 (4)
3	0	14 42 52·9	N15 33·3 (45)	12 03 05·9 (16)	11	0	15 14 25·3	N17 46·6 (39)	12 03 39·1 (5)
Fri.	6	43 52·1	15 37·8 (44)	03 07·5 (16)	Sat.	6	15 24·5	17 50·5 (39)	03 39·6 (4)
	12	44 51·2	15 42·2 (44)	03 09·1 (15)		12	16 23·6	17 54·4 (38)	03 40·0 (3)
	18	45 50·3	15 46·6 (43)	03 10·6 (15)		18	17 22·7	17 58·2 (38)	03 40·3 (4)
4	0	14 46 49·5	N15 50·9 (44)	12 03 12·1 (15)	12	0	15 18 21·9	N18 02·0 (38)	12 03 40·7 (3)
Sat.	6	47 48·6	15 55·3 (44)	03 13·6 (14)	Sun.	6	19 21·0	18 05·8 (38)	03 41·0 (2)
	12	48 47·7	15 59·7 (43)	03 15·0 (14)		12	20 20·2	18 09·6 (38)	03 41·2 (3)
	18	49 46·9	16 04·0 (43)	03 16·4 (13)		18	21 19·3	18 13·4 (37)	03 41·5 (1)
5	0	14 50 46·0	N16 08·3 (43)	12 03 17·7 (13)	13	0	15 22 18·4	N18 17·1 (37)	12 03 41·6 (2)
Sun.	6	51 45·2	16 12·6 (43)	03 19·0 (13)	Mon.	6	23 17·6	18 20·8 (37)	03 41·8 (1)
	12	52 44·3	16 16·9 (42)	03 20·3 (12)		12	24 16·7	18 24·5 (37)	03 41·9 (1)
	18	53 43·4	16 21·1 (43)	03 21·5 (12)		18	25 15·9	18 28·2 (37)	03 42·0 (0)
6	0	14 54 42·6	N16 25·4 (42)	12 03 22·7 (12)	14	0	15 26 15·0	N18 31·9 (36)	12 03 42·0 (0)
Mon.	6	55 41·7	16 29·6 (42)	03 23·9 (11)	Tues.	6	27 14·1	18 35·5 (36)	03 42·0 (0)
	12	56 40·9	16 33·8 (42)	03 25·0 (11)		12	28 13·3	18 39·1 (36)	03 42·0 (0)
	18	57 40·0	16 38·0 (42)	03 26·1 (11)		18	29 12·4	18 42·7 (36)	03 42·0 (1)
7	0	14 58 39·1	N16 42·2 (42)	12 03 27·2 (10)	15	0	15 30 11·6	N18 46·3 (36)	12 03 41·9 (2)
Tues.	6	14 59 38·3	16 46·4 (41)	03 28·2 (10)	Wed.	6	31 10·7	18 49·9 (35)	03 41·7 (1)
	12	15 00 37·4	16 50·5 (41)	03 29·2 (9)		12	32 09·8	18 53·4 (36)	03 41·6 (2)
	18	01 36·5	16 54·6 (41)	03 30·1 (9)		18	33 09·0	18 57·0 (35)	03 41·4 (3)
8	0	15 02 35·7	N16 58·7 (41)	12 03 31·0 (9)	16	0	15 34 08·1	N19 00·5 (35)	12 03 41·1 (2)
Wed.	6	03 34·8	17 02·8 (41)	03 31·9 (9)	Thur.	6	35 07·3	19 04·0 (34)	03 40·9 (3)
	12	04 34·0	17 06·9 (41)	03 32·8 (8)		12	36 06·4	19 07·4 (35)	03 40·6 (4)
	18	05 33·1	17 11·0 (41)	03 33·6 (7)		18	37 05·5	19 10·9 (34)	03 40·2 (4)
	24	15 06 32·2	N17 15·0 (40)	12 03 34·3 (7)		24	15 38 04·7	N19 14·3 (34)	12 03 39·8 (4)

Sun's SD   15′·9

SUNRISE

Date	South Latitude								0°	North Latitude							
	60°	55°	50°	45°	40°	30°	20°	10°		10°	20°	30°	40°	45°	50°	55°	60°
May	h	h	h	h	h	h	h	h	h	h	h	h	h	h	h	h	h
1	7·7	7·3	7·1	6·9	6·7	6·5	6·3	6·1	5·9	5·7	5·5	5·3	5·0	4·8	4·6	4·3	4·0
6	7·9	7·5	7·2	7·0	6·8	6·5	6·3	6·1	5·9	5·7	5·5	5·2	4·9	4·7	4·5	4·2	3·8
11	8·1	7·7	7·3	7·1	6·9	6·6	6·3	6·1	5·9	5·7	5·4	5·2	4·8	4·6	4·3	4·0	3·5
16	8·2	7·8	7·5	7·2	7·0	6·6	6·3	6·1	5·9	5·7	5·4	5·1	4·7	4·5	4·2	3·9	3·4
21	8·4	7·9	7·6	7·3	7·0	6·7	6·4	6·1	5·9	5·6	5·4	5·0	4·7	4·4	4·1	3·7	3·2
26	8·6	8·1	7·7	7·4	7·1	6·7	6·4	6·2	5·9	5·6	5·3	5·0	4·6	4·3	4·0	3·6	3·0
31	8·7	8·2	7·8	7·5	7·2	6·8	6·5	6·2	5·9	5·6	5·3	5·0	4·6	4·3	4·0	3·5	2·9

Moon's Phases: Last Quarter 4^d 07^h 16^m New Moon 12^d 10^h 45^m

UT	R	Dec	E	UT	R	Dec	E
d h	h m s	° ′	h m s	d h	h m s	° ′	h m s
17 0	15 38 04·7	N19 14·3 $_{34}$	12 03 39·8 $_4$	**25** 0	16 09 37·1	N20 52·7 $_{27}$	12 03 10·6 $_{15}$
Fri. 6	39 03·8	19 17·7 $_{34}$	03 39·4 $_4$	**Sat.** 6	10 36·3	20 55·4 $_{27}$	03 09·1 $_{14}$
12	40 03·0	19 21·1 $_{33}$	03 39·0 $_5$	12	11 35·4	20 58·1 $_{26}$	03 07·7 $_{15}$
18	41 02·1	19 24·4 $_{34}$	03 38·5 $_5$	18	12 34·5	21 00·7 $_{27}$	03 06·2 $_{16}$
18 0	15 42 01·2	N19 27·8 $_{33}$	12 03 38·0 $_5$	**26** 0	16 13 33·7	N21 03·4 $_{26}$	12 03 04·6 $_{15}$
Sat. 6	43 00·4	19 31·1 $_{33}$	03 37·5 $_6$	**Sun.** 6	14 32·8	21 06·0 $_{26}$	03 03·1 $_{16}$
12	43 59·5	19 34·4 $_{33}$	03 36·9 $_6$	12	15 32·0	21 08·6 $_{26}$	03 01·5 $_{16}$
18	44 58·7	19 37·7 $_{32}$	03 36·3 $_6$	18	16 31·1	21 11·2 $_{25}$	02 59·9 $_{17}$
19 0	15 45 57·8	N19 40·9 $_{33}$	12 03 35·7 $_7$	**27** 0	16 17 30·2	N21 13·7 $_{26}$	12 02 58·2 $_{17}$
Sun. 6	46 56·9	19 44·2 $_{32}$	03 35·0 $_7$	**Mon.** 6	18 29·4	21 16·3 $_{25}$	02 56·5 $_{17}$
12	47 56·1	19 47·4 $_{32}$	03 34·3 $_8$	12	19 28·5	21 18·8 $_{25}$	02 54·8 $_{17}$
18	48 55·2	19 50·6 $_{32}$	03 33·5 $_7$	18	20 27·7	21 21·3 $_{24}$	02 53·1 $_{18}$
20 0	15 49 54·4	N19 53·8 $_{31}$	12 03 32·8 $_8$	**28** 0	16 21 26·8	N21 23·7 $_{25}$	12 02 51·3 $_{18}$
Mon. 6	50 53·5	19 56·9 $_{31}$	03 32·0 $_8$	**Tues.** 6	22 25·9	21 26·2 $_{24}$	02 49·5 $_{18}$
12	51 52·6	20 00·0 $_{31}$	03 31·1 $_9$	12	23 25·1	21 28·6 $_{24}$	02 47·7 $_{19}$
18	52 51·8	20 03·1 $_{31}$	03 30·2 $_9$	18	24 24·2	21 31·0 $_{23}$	02 45·8 $_{19}$
21 0	15 53 50·9	N20 06·2 $_{31}$	12 03 29·3 $_9$	**29** 0	16 25 23·4	N21 33·3 $_{24}$	12 02 43·9 $_{19}$
Tues. 6	54 50·1	20 09·3 $_{30}$	03 28·4 $_{10}$	**Wed.** 6	26 22·5	21 35·7 $_{23}$	02 42·0 $_{19}$
12	55 49·2	20 12·3 $_{31}$	03 27·4 $_{10}$	12	27 21·7	21 38·0 $_{23}$	02 40·1 $_{20}$
18	56 48·3	20 15·4 $_{30}$	03 26·4 $_{10}$	18	28 20·8	21 40·3 $_{23}$	02 38·1 $_{20}$
22 0	15 57 47·5	N20 18·4 $_{29}$	12 03 25·4 $_{11}$	**30** 0	16 29 19·9	N21 42·6 $_{22}$	12 02 36·1 $_{20}$
Wed. 6	58 46·6	20 21·3 $_{30}$	03 24·3 $_{11}$	**Thur.** 6	30 19·1	21 44·8 $_{23}$	02 34·1 $_{21}$
12	15 59 45·7	20 24·3 $_{29}$	03 23·2 $_{11}$	12	31 18·2	21 47·1 $_{22}$	02 32·0 $_{21}$
18	16 00 44·9	20 27·2 $_{30}$	03 22·1 $_{11}$	18	32 17·4	21 49·3 $_{21}$	02 29·9 $_{21}$
23 0	16 01 44·0	N20 30·2 $_{28}$	12 03 21·0 $_{12}$	**31** 0	16 33 16·5	N21 51·4 $_{22}$	12 02 27·8 $_{21}$
Thur. 6	02 43·2	20 33·0 $_{29}$	03 19·8 $_{12}$	**Fri.** 6	34 15·6	21 53·6 $_{21}$	02 25·7 $_{22}$
12	03 42·3	20 35·9 $_{29}$	03 18·6 $_{13}$	12	35 14·8	21 55·7 $_{21}$	02 23·5 $_{22}$
18	04 41·4	20 38·8 $_{28}$	03 17·3 $_{13}$	18	36 13·9	21 57·8 $_{21}$	02 21·3 $_{22}$
24 0	16 05 40·6	N20 41·6 $_{28}$	12 03 16·0 $_{13}$	**32** 0	16 37 13·1	N21 59·9 $_{21}$	12 02 19·1 $_{22}$
Fri. 6	06 39·7	20 44·4 $_{28}$	03 14·7 $_{13}$	**Sat.** 6	38 12·2	22 02·0 $_{20}$	02 16·9 $_{23}$
12	07 38·8	20 47·2 $_{27}$	03 13·4 $_{14}$	12	39 11·3	22 04·0 $_{21}$	02 14·6 $_{23}$
18	08 38·0	20 49·9 $_{28}$	03 12·0 $_{14}$	18	40 10·5	22 06·1 $_{19}$	02 12·3 $_{23}$
24	16 09 37·1	N20 52·7	12 03 10·6	24	16 41 09·6	N22 08·0	12 02 10·0

Sun's SD   15·8

SUNSET

Date	South Latitude								0°	North Latitude							
	60°	55°	50°	45°	40°	30°	20°	10°		10°	20°	30°	40°	45°	50°	55°	60°
May	h	h	h	h	h	h	h	h	h	h	h	h	h	h	h	h	h
1	16·2	16·6	16·8	17·0	17·1	17·4	17·6	17·8	18·0	18·2	18·4	18·6	18·9	19·1	19·3	19·6	19·9
6	16·0	16·4	16·6	16·9	17·1	17·4	17·6	17·8	18·0	18·2	18·4	18·7	19·0	19·2	19·4	19·7	20·1
11	15·8	16·2	16·5	16·8	17·0	17·3	17·6	17·8	18·0	18·2	18·4	18·7	19·1	19·3	19·6	19·9	20·4
16	15·6	16·1	16·4	16·7	16·9	17·2	17·5	17·8	18·0	18·2	18·5	18·8	19·1	19·4	19·7	20·0	20·6
21	15·4	15·9	16·3	16·6	16·8	17·2	17·5	17·7	18·0	18·2	18·5	18·8	19·2	19·5	19·8	20·2	20·7
26	15·3	15·8	16·2	16·5	16·8	17·2	17·5	17·7	18·0	18·3	18·6	18·9	19·3	19·6	19·9	20·3	20·9
31	15·2	15·7	16·1	16·5	16·7	17·1	17·5	17·7	18·0	18·3	18·6	18·9	19·4	19·6	20·0	20·4	21·1

Moon's Phases: First Quarter 19^d 19^h 42^m Full Moon 26^d 11^h 51^m

SUN – JUNE, 2002

UT		R	Dec	E	UT		R	Dec	E
d	h	h m s	° '	h m s	d	h	h m s	° '	h m s
1	**0**	16 37 13·1	N 21 59·9 ₂₁	12 02 19·1 ₂₂	**9**	**0**	17 08 45·5	N 22 53·9 ₁₂	12 00 55·7 ₂₉
Sat.	6	38 12·2	22 02·0 ₂₀	02 16·9 ₂₃	Sun.	6	09 44·6	22 55·1 ₁₃	00 52·8 ₃₀
	12	39 11·3	22 04·0 ₂₁	02 14·6 ₂₃		12	10 43·8	22 56·4 ₁₂	00 49·8 ₃₀
	18	40 10·5	22 06·1 ₁₉	02 12·3 ₂₃		18	11 42·9	22 57·6 ₁₂	00 46·8 ₂₉
2	**0**	16 41 09·6	N 22 08·0 ₂₀	12 02 10·0 ₂₄	**10**	**0**	17 12 42·0	N 22 58·8 ₁₂	12 00 43·9 ₃₀
Sun.	6	42 08·7	22 10·0 ₂₀	02 07·6 ₂₃	Mon.	6	13 41·2	23 00·0 ₁₁	00 40·9 ₃₁
	12	43 07·9	22 12·0 ₁₉	02 05·3 ₂₄		12	14 40·3	23 01·1 ₁₂	00 37·8 ₃₀
	18	44 07·0	22 13·9 ₁₉	02 02·9 ₂₅		18	15 39·5	23 02·3 ₁₁	00 34·8 ₃₀
3	**0**	16 45 06·2	N 22 15·8 ₁₈	12 02 00·4 ₂₄	**11**	**0**	17 16 38·6	N 23 03·4 ₁₀	12 00 31·8 ₃₁
Mon.	6	46 05·3	22 17·6 ₁₉	01 58·0 ₂₅	Tues.	6	17 37·7	23 04·4 ₁₁	00 28·7 ₃₀
	12	47 04·4	22 19·5 ₁₈	01 55·5 ₂₅		12	18 36·9	23 05·5 ₁₀	00 25·7 ₃₁
	18	48 03·6	22 21·3 ₁₈	01 53·0 ₂₅		18	19 36·0	23 06·5 ₁₀	00 22·6 ₃₁
4	**0**	16 49 02·7	N 22 23·1 ₁₈	12 01 50·5 ₂₅	**12**	**0**	17 20 35·2	N 23 07·5 ₁₀	12 00 19·5 ₃₁
Tues.	6	50 01·9	22 24·9 ₁₇	01 48·0 ₂₆	Wed.	6	21 34·3	23 08·5 ₉	00 16·4 ₃₁
	12	51 01·0	22 26·6 ₁₇	01 45·4 ₂₆		12	22 33·4	23 09·4 ₉	00 13·3 ₃₂
	18	52 00·1	22 28·3 ₁₇	01 42·8 ₂₆		18	23 32·6	23 10·3 ₉	00 10·1 ₃₂
5	**0**	16 52 59·3	N 22 30·0 ₁₇	12 01 40·2 ₂₆	**13**	**0**	17 24 31·7	N 23 11·2 ₉	12 00 07·0 ₃₂
Wed.	6	53 58·4	22 31·7 ₁₇	01 37·6 ₂₇	Thur.	6	25 30·9	23 12·1 ₉	00 03·8 ₃₁
	12	54 57·5	22 33·4 ₁₆	01 34·9 ₂₆		12	26 30·0	23 13·0 ₈	12 00 00·7 ₃₂
	18	55 56·7	22 35·0 ₁₆	01 32·3 ₂₇		18	27 29·2	23 13·8 ₈	11 59 57·5 ₃₁
6	**0**	16 56 55·8	N 22 36·6 ₁₆	12 01 29·6 ₂₈	**14**	**0**	17 28 28·3	N 23 14·6 ₇	11 59 54·4 ₃₂
Thur.	6	57 55·0	22 38·2 ₁₅	01 26·8 ₂₇	Fri.	6	29 27·4	23 15·3 ₈	59 51·2 ₃₂
	12	58 54·1	22 39·7 ₁₅	01 24·1 ₂₇		12	30 26·6	23 16·1 ₇	59 48·0 ₃₂
	18	16 59 53·2	22 41·2 ₁₅	01 21·4 ₂₈		18	31 25·7	23 16·8 ₇	59 44·8 ₃₂
7	**0**	17 00 52·4	N 22 42·7 ₁₅	12 01 18·6 ₂₈	**15**	**0**	17 32 24·9	N 23 17·5 ₆	11 59 41·6 ₃₂
Fri.	6	01 51·5	22 44·2 ₁₅	01 15·8 ₂₈	Sat.	6	33 24·0	23 18·1 ₇	59 38·4 ₃₃
	12	02 50·6	22 45·7 ₁₄	01 13·0 ₂₉		12	34 23·1	23 18·8 ₆	59 35·1 ₃₂
	18	03 49·8	22 47·1 ₁₄	01 10·1 ₂₈		18	35 22·3	23 19·4 ₆	59 31·9 ₃₂
8	**0**	17 04 48·9	N 22 48·5 ₁₄	12 01 07·3 ₂₉	**16**	**0**	17 36 21·4	N 23 20·0 ₅	11 59 28·7 ₃₂
Sat.	6	05 48·1	22 49·9 ₁₃	01 04·4 ₂₉	Sun.	6	37 20·6	23 20·5 ₆	59 25·5 ₃₃
	12	06 47·2	22 51·2 ₁₄	01 01·5 ₂₉		12	38 19·7	23 21·1 ₅	59 22·2 ₃₃
	18	07 46·3	22 52·6 ₁₃	00 58·6 ₂₉		18	39 18·8	23 21·6 ₅	59 19·0 ₃₂
	24	17 08 45·5	N 22 53·9	12 00 55·7 ₂₉		24	17 40 18·0	N 23 22·1 ₅	11 59 15·7 ₃₃

Sun's SD   15·8

SUNRISE

Date	South Latitude								0°	North Latitude							
	60°	55°	50°	45°	40°	30°	20°	10°		10°	20°	30°	40°	45°	50°	55°	60°
June	h	h	h	h	h	h	h	h	h	h	h	h	h	h	h	h	h
1	8·8	8·2	7·8	7·5	7·2	6·8	6·5	6·2	5·9	5·6	5·3	5·0	4·5	4·3	3·9	3·5	2·8
6	8·9	8·3	7·8	7·5	7·2	6·8	6·5	6·2	5·9	5·6	5·3	5·0	4·5	4·2	3·9	3·4	2·7
11	9·0	8·3	7·9	7·6	7·3	6·9	6·5	6·2	5·9	5·7	5·3	5·0	4·5	4·2	3·9	3·4	2·6
16	9·1	8·4	8·0	7·6	7·3	6·9	6·5	6·2	6·0	5·7	5·3	5·0	4·5	4·2	3·8	3·3	2·6
21	9·1	8·4	8·0	7·7	7·4	6·9	6·6	6·2	6·0	5·7	5·3	5·0	4·5	4·2	3·8	3·3	2·6
26	9·1	8·4	8·0	7·7	7·4	6·9	6·6	6·3	6·0	5·7	5·4	5·0	4·5	4·2	3·9	3·4	2·6
31	9·1	8·4	8·0	7·7	7·4	6·9	6·6	6·3	6·0	5·7	5·4	5·0	4·6	4·3	3·9	3·4	2·7

Moon's Phases: Last Quarter 3ᵈ 00ʰ 05ᵐ   New Moon 10ᵈ 23ʰ 46ᵐ

UT	R	Dec	E	UT	R	Dec	E
d h	h m s	° ′	h m s	d h	h m s	° ′	h m s
17 Mon. 0	17 40 18·0	N23 22·1$_4$	11 59 15·7$_{32}$	**25** Tues. 0	18 11 50·4	N23 23·9$_4$	11 57 32·0$_{32}$
6	41 17·1	23 22·5$_5$	59 12·5$_{33}$	6	12 49·6	23 23·5$_3$	57 28·8$_{32}$
12	42 16·2	23 23·0$_5$	59 09·2$_{33}$	12	13 48·7	23 23·2$_5$	57 25·6$_{31}$
18	43 15·4	23 23·4$_4$	59 06·0$_{33}$	18	14 47·9	23 22·7$_4$	57 22·5$_{32}$
18 Tues. 0	17 44 14·5	N23 23·8$_3$	11 59 02·7$_{32}$	**26** Wed. 0	18 15 47·0	N23 22·3$_5$	11 57 19·3$_{31}$
6	45 13·7	23 24·1$_3$	58 59·5$_{33}$	6	16 46·1	23 21·8$_5$	57 16·2$_{32}$
12	46 12·8	23 24·4$_3$	58 56·2$_{33}$	12	17 45·3	23 21·3$_5$	57 13·0$_{31}$
18	47 11·9	23 24·7$_3$	58 52·9$_{32}$	18	18 44·4	23 20·8$_6$	57 09·9$_{31}$
19 Wed. 0	17 48 11·1	N23 25·0$_3$	11 58 49·7$_{33}$	**27** Thur. 0	18 19 43·6	N23 20·2$_5$	11 57 06·8$_{31}$
6	49 10·2	23 25·3$_2$	58 46·4$_{33}$	6	20 42·7	23 19·7$_6$	57 03·7$_{31}$
12	50 09·3	23 25·5$_2$	58 43·1$_{32}$	12	21 41·8	23 19·1$_6$	57 00·6$_{31}$
18	51 08·5	23 25·7$_2$	58 39·9$_{33}$	18	22 41·0	23 18·5$_7$	56 57·5$_{31}$
20 Thur. 0	17 52 07·6	N23 25·9$_1$	11 58 36·6$_{32}$	**28** Fri. 0	18 23 40·1	N23 17·8$_7$	11 56 54·4$_{31}$
6	53 06·8	23 26·0$_1$	58 33·4$_{33}$	6	24 39·3	23 17·1$_7$	56 51·3$_{30}$
12	54 05·9	23 26·1$_1$	58 30·1$_{33}$	12	25 38·4	23 16·4$_7$	56 48·3$_{31}$
18	55 05·0	23 26·2$_1$	58 26·8$_{32}$	18	26 37·5	23 15·7$_8$	56 45·2$_{30}$
21 Fri. 0	17 56 04·2	N23 26·3$_0$	11 58 23·6$_{33}$	**29** Sat. 0	18 27 36·7	N23 14·9$_7$	11 56 42·2$_{30}$
6	57 03·3	23 26·3$_1$	58 20·3$_{32}$	6	28 35·8	23 14·2$_8$	56 39·2$_{30}$
12	58 02·5	23 26·4$_0$	58 17·1$_{33}$	12	29 35·0	23 13·4$_9$	56 36·2$_{30}$
18	17 59 01·6	23 26·4$_1$	58 13·8$_{32}$	18	30 34·1	23 12·5$_8$	56 33·2$_{30}$
22 Sat. 0	18 00 00·7	N23 26·3$_0$	11 58 10·6$_{33}$	**30** Sun. 0	18 31 33·2	N23 11·7$_9$	11 56 30·2$_{30}$
6	00 59·9	23 26·3$_1$	58 07·3$_{32}$	6	32 32·4	23 10·8$_9$	56 27·2$_{30}$
12	01 59·0	23 26·2$_1$	58 04·1$_{32}$	12	33 31·5	23 09·9$_9$	56 24·2$_{29}$
18	02 58·2	23 26·1$_2$	58 00·9$_{33}$	18	34 30·6	23 09·0$_{10}$	56 21·3$_{29}$
23 Sun. 0	18 03 57·3	N23 25·9$_1$	11 57 57·6$_{32}$	**31** Mon. 0	18 35 29·8	N23 08·0$_{10}$	11 56 18·4$_{30}$
6	04 56·4	23 25·8$_2$	57 54·4$_{32}$	6	36 28·9	23 07·0$_{10}$	56 15·4$_{29}$
12	05 55·6	23 25·6$_2$	57 51·2$_{32}$	12	37 28·1	23 06·0$_{10}$	56 12·5$_{29}$
18	06 54·7	23 25·4$_3$	57 48·0$_{32}$	18	38 27·2	23 05·0$_{11}$	56 09·6$_{28}$
24 Mon. 0	18 07 53·9	N23 25·1$_2$	11 57 44·8$_{32}$	24	18 39 26·3	N23 03·9	11 56 06·8
6	08 53·0	23 24·9$_3$	57 41·6$_{32}$				
12	09 52·1	23 24·6$_3$	57 38·4$_{32}$				
18	10 51·3	23 24·3$_4$	57 35·2$_{32}$				
24	18 11 50·4	N23 23·9	11 57 32·0				

Sun's SD 15′·8

SUNSET

Date	South Latitude									North Latitude							
	60°	55°	50°	45°	40°	30°	20°	10°	0°	10°	20°	30°	40°	45°	50°	55°	60°
June	h	h	h	h	h	h	h	h	h	h	h	h	h	h	h	h	h
1	15·2	15·7	16·1	16·5	16·7	17·1	17·5	17·7	18·0	18·3	18·6	18·9	19·4	19·6	20·0	20·4	21·1
6	15·1	15·7	16·1	16·4	16·7	17·1	17·4	17·7	18·0	18·3	18·6	19·0	19·4	19·7	20·1	20·6	21·2
11	15·0	15·6	16·1	16·4	16·7	17·1	17·5	17·8	18·1	18·3	18·6	19·0	19·5	19·8	20·1	20·6	21·4
16	15·0	15·6	16·1	16·4	16·7	17·1	17·5	17·8	18·1	18·4	18·7	19·1	19·5	19·8	20·2	20·7	21·4
21	15·0	15·6	16·1	16·4	16·7	17·1	17·5	17·8	18·1	18·4	18·7	19·1	19·5	19·8	20·2	20·7	21·5
26	15·0	15·6	16·1	16·4	16·7	17·1	17·5	17·8	18·1	18·4	18·7	19·1	19·6	19·9	20·2	20·7	21·5
31	15·1	15·7	16·1	16·5	16·7	17·2	17·5	17·8	18·1	18·4	18·7	19·1	19·6	19·8	20·2	20·7	21·4

Moon's Phases: First Quarter 18^d 00^h 29^m Full Moon 24^d 21^h 42^m

SUN – JULY, 2002

	UT	R	Dec	E
d	h	h m s	° ′	h m s
1 **Mon.**	**0**	18 35 29·8	N23 08·0 $_{10}$	11 56 18·4 $_{30}$
	6	36 28·9	23 07·0 $_{10}$	56 15·4 $_{29}$
	12	37 28·1	23 06·0 $_{10}$	56 12·5 $_{29}$
	18	38 27·2	23 05·0 $_{11}$	56 09·6 $_{28}$
2 **Tues.**	**0**	18 39 26·3	N23 03·9 $_{10}$	11 56 06·8 $_{29}$
	6	40 25·5	23 02·9 $_{11}$	56 03·9 $_{28}$
	12	41 24·6	23 01·8 $_{12}$	56 01·1 $_{29}$
	18	42 23·7	23 00·6 $_{11}$	55 58·2 $_{28}$
3 **Wed.**	**0**	18 43 22·9	N22 59·5 $_{12}$	11 55 55·4 $_{28}$
	6	44 22·0	22 58·3 $_{12}$	55 52·6 $_{27}$
	12	45 21·2	22 57·1 $_{13}$	55 49·9 $_{28}$
	18	46 20·3	22 55·8 $_{12}$	55 47·1 $_{27}$
4 **Thur.**	**0**	18 47 19·4	N22 54·6 $_{13}$	11 55 44·4 $_{27}$
	6	48 18·6	22 53·3 $_{13}$	55 41·7 $_{27}$
	12	49 17·7	22 52·0 $_{13}$	55 39·0 $_{27}$
	18	50 16·9	22 50·7 $_{14}$	55 36·3 $_{27}$
5 **Fri.**	**0**	18 51 16·0	N22 49·3 $_{14}$	11 55 33·6 $_{26}$
	6	52 15·1	22 47·9 $_{14}$	55 31·0 $_{27}$
	12	53 14·3	22 46·5 $_{14}$	55 28·3 $_{26}$
	18	54 13·4	22 45·1 $_{15}$	55 25·7 $_{25}$
6 **Sat.**	**0**	18 55 12·5	N22 43·6 $_{14}$	11 55 23·2 $_{26}$
	6	56 11·7	22 42·2 $_{16}$	55 20·6 $_{25}$
	12	57 10·8	22 40·6 $_{15}$	55 18·1 $_{26}$
	18	58 10·0	22 39·1 $_{15}$	55 15·5 $_{25}$
7 **Sun.**	**0**	18 59 09·1	N22 37·6 $_{16}$	11 55 13·0 $_{24}$
	6	19 00 08·2	22 36·0 $_{16}$	55 10·6 $_{25}$
	12	01 07·4	22 34·4 $_{16}$	55 08·1 $_{24}$
	18	02 06·5	22 32·8 $_{17}$	55 05·7 $_{24}$
8 **Mon.**	**0**	19 03 05·7	N22 31·1 $_{17}$	11 55 03·3 $_{24}$
	6	04 04·8	22 29·4 $_{17}$	55 00·9 $_{24}$
	12	05 03·9	22 27·7 $_{17}$	54 58·5 $_{23}$
	18	06 03·1	22 26·0 $_{18}$	54 56·2 $_{23}$
	24	19 07 02·2	N22 24·2	11 54 53·9
9 **Tues.**	**0**	19 07 02·2	N22 24·2 $_{17}$	11 54 53·9 $_{23}$
	6	08 01·4	22 22·5 $_{18}$	54 51·6 $_{23}$
	12	09 00·5	22 20·7 $_{18}$	54 49·3 $_{22}$
	18	09 59·7	22 18·9 $_{19}$	54 47·1 $_{22}$
10 **Wed.**	**0**	19 10 58·8	N22 17·0 $_{19}$	11 54 44·9 $_{22}$
	6	11 57·9	22 15·1 $_{19}$	54 42·7 $_{21}$
	12	12 57·1	22 13·2 $_{19}$	54 40·6 $_{22}$
	18	13 56·2	22 11·3 $_{19}$	54 38·4 $_{21}$
11 **Thur.**	**0**	19 14 55·4	N22 09·4 $_{20}$	11 54 36·3 $_{20}$
	6	15 54·5	22 07·4 $_{20}$	54 34·3 $_{21}$
	12	16 53·6	22 05·4 $_{20}$	54 32·2 $_{20}$
	18	17 52·8	22 03·4 $_{20}$	54 30·2 $_{20}$
12 **Fri.**	**0**	19 18 51·9	N22 01·4 $_{21}$	11 54 28·2 $_{19}$
	6	19 51·1	21 59·3 $_{21}$	54 26·3 $_{20}$
	12	20 50·2	21 57·2 $_{21}$	54 24·3 $_{19}$
	18	21 49·3	21 55·1 $_{21}$	54 22·4 $_{18}$
13 **Sat.**	**0**	19 22 48·5	N21 53·0 $_{22}$	11 54 20·6 $_{19}$
	6	23 47·6	21 50·8 $_{21}$	54 18·7 $_{18}$
	12	24 46·8	21 48·7 $_{22}$	54 16·9 $_{18}$
	18	25 45·9	21 46·5 $_{23}$	54 15·1 $_{17}$
14 **Sun.**	**0**	19 26 45·0	N21 44·2 $_{22}$	11 54 13·4 $_{17}$
	6	27 44·2	21 42·0 $_{23}$	54 11·7 $_{17}$
	12	28 43·3	21 39·7 $_{23}$	54 10·0 $_{17}$
	18	29 42·4	21 37·4 $_{23}$	54 08·3 $_{16}$
15 **Mon.**	**0**	19 30 41·6	N21 35·1 $_{23}$	11 54 06·7 $_{16}$
	6	31 40·7	21 32·8 $_{24}$	54 05·1 $_{15}$
	12	32 39·9	21 30·4 $_{24}$	54 03·6 $_{15}$
	18	33 39·0	21 28·0 $_{24}$	54 02·1 $_{15}$
16 **Tues.**	**0**	19 34 38·1	N21 25·6 $_{24}$	11 54 00·6 $_{15}$
	6	35 37·3	21 23·2 $_{25}$	53 59·1 $_{14}$
	12	36 36·4	21 20·7 $_{24}$	53 57·7 $_{14}$
	18	37 35·6	21 18·3 $_{25}$	53 56·3 $_{13}$
	24	19 38 34·7	N21 15·8 $_{25}$	11 53 55·0

Sun's SD   15′·8

SUNRISE

Date	South Latitude								0°	North Latitude							
	60°	55°	50°	45°	40°	30°	20°	10°	0°	10°	20°	30°	40°	45°	50°	55°	60°
July	h	h	h	h	h	h	h	h	h	h	h	h	h	h	h	h	h
1	9·1	8·4	8·0	7·7	7·4	6·9	6·6	6·3	6·0	5·7	5·4	5·0	4·6	4·3	3·9	3·4	2·7
6	9·0	8·4	8·0	7·6	7·4	6·9	6·6	6·3	6·0	5·7	5·4	5·1	4·6	4·3	4·0	3·5	2·8
11	8·9	8·3	7·9	7·6	7·3	6·9	6·6	6·3	6·0	5·7	5·5	5·1	4·7	4·4	4·0	3·6	2·9
16	8·8	8·2	7·8	7·5	7·3	6·9	6·6	6·3	6·0	5·8	5·5	5·2	4·7	4·5	4·1	3·7	3·1
21	8·7	8·1	7·8	7·5	7·2	6·9	6·6	6·3	6·0	5·8	5·5	5·2	4·8	4·5	4·2	3·8	3·2
26	8·5	8·0	7·7	7·4	7·2	6·8	6·5	6·3	6·0	5·8	5·5	5·2	4·9	4·6	4·3	4·0	3·4
31	8·3	7·9	7·6	7·3	7·1	6·8	6·5	6·3	6·0	5·8	5·6	5·3	5·0	4·7	4·5	4·1	3·6

Moon's Phases: Last Quarter 2^d 17^h 19^m New Moon 10^d 10^h 26^m

UT		R	Dec	E	UT		R	Dec	E
d	h	h m s	° ′	h m s	d	h	h m s	° ′	h m s
17	0	19 38 34·7	N21 15·8 ₍₂₆₎	11 53 55·0 ₍₁₄₎	**25**	0	20 10 07·2	N19 44·2 ₍₃₂₎	11 53 30·3 ₍₂₎
Wed.	6	39 33·8	21 13·2 ₍₂₅₎	53 53·6 ₍₁₃₎	**Thur.**	6	11 06·3	19 41·0 ₍₃₂₎	53 30·1 ₍₁₎
	12	40 33·0	21 10·7 ₍₂₆₎	53 52·3 ₍₁₂₎		12	12 05·4	19 37·8 ₍₃₃₎	53 30·0 ₍₁₎
	18	41 32·1	21 08·1 ₍₂₆₎	53 51·1 ₍₁₂₎		18	13 04·6	19 34·5 ₍₃₂₎	53 29·9 ₍₁₎
18	0	19 42 31·2	N21 05·5 ₍₂₆₎	11 53 49·9 ₍₁₂₎	**26**	0	20 14 03·7	N19 31·3 ₍₃₃₎	11 53 29·8 ₍₁₎
Thur.	6	43 30·4	21 02·9 ₍₂₆₎	53 48·7 ₍₁₂₎	**Fri.**	6	15 02·9	19 28·0 ₍₃₃₎	53 29·8 ₍₀₎
	12	44 29·5	21 00·3 ₍₂₇₎	53 47·5 ₍₁₁₎		12	16 02·0	19 24·7 ₍₃₄₎	53 29·8 ₍₀₎
	18	45 28·7	20 57·6 ₍₂₆₎	53 46·4 ₍₁₀₎		18	17 01·1	19 21·3 ₍₃₃₎	53 29·9 ₍₀₎
19	0	19 46 27·8	N20 55·0 ₍₂₇₎	11 53 45·4 ₍₁₁₎	**27**	0	20 18 00·3	N19 18·0 ₍₃₄₎	11 53 29·9 ₍₂₎
Fri.	6	47 26·9	20 52·3 ₍₂₈₎	53 44·3 ₍₁₀₎	**Sat.**	6	18 59·4	19 14·6 ₍₃₄₎	53 30·1 ₍₁₎
	12	48 26·1	20 49·5 ₍₂₇₎	53 43·3 ₍₁₀₎		12	19 58·6	19 11·2 ₍₃₄₎	53 30·2 ₍₂₎
	18	49 25·2	20 46·8 ₍₂₈₎	53 42·3 ₍₉₎		18	20 57·7	19 07·8 ₍₃₄₎	53 30·4 ₍₃₎
20	0	19 50 24·4	N20 44·0 ₍₂₈₎	11 53 41·4 ₍₉₎	**28**	0	20 21 56·8	N19 04·4 ₍₃₅₎	11 53 30·7 ₍₂₎
Sat.	6	51 23·5	20 41·2 ₍₂₈₎	53 40·5 ₍₉₎	**Sun.**	6	22 56·0	19 00·9 ₍₃₄₎	53 30·9 ₍₃₎
	12	52 22·6	20 38·4 ₍₂₈₎	53 39·6 ₍₈₎		12	23 55·1	18 57·5 ₍₃₅₎	53 31·2 ₍₄₎
	18	53 21·8	20 35·6 ₍₂₉₎	53 38·8 ₍₈₎		18	24 54·2	18 54·0 ₍₃₅₎	53 31·6 ₍₄₎
21	0	19 54 20·9	N20 32·7 ₍₂₈₎	11 53 38·0 ₍₇₎	**29**	0	20 25 53·4	N18 50·5 ₍₃₆₎	11 53 32·0 ₍₄₎
Sun.	6	55 20·1	20 29·9 ₍₂₉₎	53 37·3 ₍₇₎	**Mon.**	6	26 52·5	18 46·9 ₍₃₅₎	53 32·4 ₍₄₎
	12	56 19·2	20 27·0 ₍₂₉₎	53 36·5 ₍₆₎		12	27 51·7	18 43·4 ₍₃₆₎	53 32·8 ₍₄₎
	18	57 18·3	20 24·1 ₍₃₀₎	53 35·9 ₍₇₎		18	28 50·8	18 39·8 ₍₃₆₎	53 33·3 ₍₆₎
22	0	19 58 17·5	N20 21·1 ₍₂₉₎	11 53 35·2 ₍₆₎	**30**	0	20 29 49·9	N18 36·2 ₍₃₆₎	11 53 33·9 ₍₅₎
Mon.	6	19 59 16·6	20 18·2 ₍₃₀₎	53 34·6 ₍₆₎	**Tues.**	6	30 49·1	18 32·6 ₍₃₆₎	53 34·4 ₍₆₎
	12	20 00 15·8	20 15·2 ₍₃₀₎	53 34·0 ₍₅₎		12	31 48·2	18 29·0 ₍₃₆₎	53 35·0 ₍₇₎
	18	01 14·9	20 12·2 ₍₃₀₎	53 33·5 ₍₅₎		18	32 47·3	18 25·4 ₍₃₇₎	53 35·7 ₍₆₎
23	0	20 02 14·0	N20 09·2 ₍₃₁₎	11 53 33·0 ₍₅₎	**31**	0	20 33 46·5	N18 21·7 ₍₃₇₎	11 53 36·3 ₍₇₎
Tues.	6	03 13·2	20 06·1 ₍₃₁₎	53 32·5 ₍₄₎	**Wed.**	6	34 45·6	18 18·0 ₍₃₇₎	53 37·0 ₍₈₎
	12	04 12·3	20 03·0 ₍₃₀₎	53 32·1 ₍₄₎		12	35 44·8	18 14·3 ₍₃₇₎	53 37·8 ₍₈₎
	18	05 11·5	20 00·0 ₍₃₁₎	53 31·7 ₍₃₎		18	36 43·9	18 10·6 ₍₃₇₎	53 38·6 ₍₈₎
24	0	20 06 10·6	N19 56·9 ₍₃₂₎	11 53 31·4 ₍₄₎	**32**	0	20 37 43·0	N18 06·9 ₍₃₈₎	11 53 39·4 ₍₉₎
Wed.	6	07 09·7	19 53·7 ₍₃₁₎	53 31·0 ₍₃₎	**Thur.**	6	38 42·2	18 03·1 ₍₃₇₎	53 40·3 ₍₉₎
	12	08 08·9	19 50·6 ₍₃₂₎	53 30·7 ₍₂₎		12	39 41·3	17 59·4 ₍₃₈₎	53 41·2 ₍₉₎
	18	09 08·0	19 47·4 ₍₃₂₎	53 30·5 ₍₂₎		18	40 40·4	17 55·6 ₍₃₈₎	53 42·1 ₍₁₀₎
	24	20 10 07·2	N19 44·2	11 53 30·3		24	20 41 39·6	N17 51·8 ₍₃₈₎	11 53 43·1

Sun's SD   15′·8

SUNSET

Date	South Latitude								0°	North Latitude							
	60°	55°	50°	45°	40°	30°	20°	10°		10°	20°	30°	40°	45°	50°	55°	60°
July	h	h	h	h	h	h	h	h	h	h	h	h	h	h	h	h	h
1	15·1	15·7	16·1	16·5	16·7	17·2	17·5	17·8	18·1	18·4	18·7	19·1	19·6	19·8	20·2	20·7	21·4
6	15·2	15·8	16·2	16·5	16·8	17·2	17·6	17·9	18·1	18·4	18·7	19·1	19·5	19·8	20·2	20·6	21·4
11	15·3	15·8	16·3	16·6	16·9	17·2	17·6	17·9	18·1	18·4	18·7	19·1	19·5	19·8	20·1	20·6	21·2
16	15·4	16·0	16·4	16·6	16·9	17·3	17·6	17·9	18·2	18·4	18·7	19·0	19·4	19·7	20·1	20·5	21·1
21	15·6	16·1	16·4	16·7	17·0	17·4	17·6	17·9	18·2	18·4	18·7	19·0	19·4	19·6	20·0	20·4	20·9
26	15·7	16·2	16·6	16·8	17·0	17·4	17·7	17·9	18·2	18·4	18·7	19·0	19·3	19·6	19·9	20·2	20·8
31	15·9	16·4	16·7	16·9	17·1	17·4	17·7	17·9	18·2	18·4	18·6	18·9	19·2	19·5	19·7	20·1	20·6

Moon's Phases: First Quarter 17ᵈ 04ʰ 47ᵐ   Full Moon 24ᵈ 09ʰ 07ᵐ   Last Quarter 32ᵈ 10ʰ 22ᵐ

SUN – AUGUST, 2002

UT d	h	R h m s	Dec ° '	E h m s		UT d	h	R h m s	Dec ° '	E h m s
1	0	20 37 43·0	N18 06·9$_{38}$	11 53 39·4$_9$		9	0	21 09 15·5	N15 57·9$_{43}$	11 54 25·1$_{21}$
Thur.	6	38 42·2	18 03·1$_{37}$	53 40·3$_9$		Fri.	6	10 14·6	15 53·6$_{43}$	54 27·2$_{21}$
	12	39 41·3	17 59·4$_{38}$	53 41·2$_9$			12	11 13·8	15 49·3$_{43}$	54 29·3$_{21}$
	18	40 40·4	17 55·6$_{38}$	53 42·1$_{10}$			18	12 12·9	15 45·0$_{44}$	54 31·4$_{21}$
2	0	20 41 39·6	N17 51·8$_{39}$	11 53 43·1$_{10}$		10	0	21 13 12·1	N15 40·6$_{44}$	11 54 33·5$_{22}$
Fri.	6	42 38·7	17 47·9$_{38}$	53 44·1$_{10}$		Sat.	6	14 11·2	15 36·2$_{43}$	54 35·7$_{22}$
	12	43 37·9	17 44·1$_{39}$	53 45·1$_{11}$			12	15 10·3	15 31·9$_{44}$	54 37·9$_{23}$
	18	44 37·0	17 40·2$_{39}$	53 46·2$_{11}$			18	16 09·5	15 27·5$_{45}$	54 40·2$_{22}$
3	0	20 45 36·1	N17 36·3$_{39}$	11 53 47·3$_{12}$		11	0	21 17 08·6	N15 23·0$_{44}$	11 54 42·4$_{24}$
Sat.	6	46 35·3	17 32·4$_{39}$	53 48·5$_{12}$		Sun.	6	18 07·7	15 18·6$_{44}$	54 44·8$_{23}$
	12	47 34·4	17 28·5$_{39}$	53 49·7$_{12}$			12	19 06·9	15 14·2$_{45}$	54 47·1$_{24}$
	18	48 33·6	17 24·6$_{40}$	53 50·9$_{13}$			18	20 06·0	15 09·7$_{45}$	54 49·5$_{25}$
4	0	20 49 32·7	N17 20·6$_{40}$	11 53 52·2$_{13}$		12	0	21 21 05·2	N15 05·2$_{45}$	11 54 52·0$_{24}$
Sun.	6	50 31·8	17 16·6$_{39}$	53 53·5$_{13}$		Mon.	6	22 04·3	15 00·7$_{45}$	54 54·4$_{25}$
	12	51 31·0	17 12·7$_{40}$	53 54·8$_{14}$			12	23 03·4	14 56·2$_{45}$	54 56·9$_{26}$
	18	52 30·1	17 08·7$_{41}$	53 56·2$_{14}$			18	24 02·6	14 51·7$_{46}$	54 59·5$_{26}$
5	0	20 53 29·3	N17 04·6$_{40}$	11 53 57·6$_{14}$		13	0	21 25 01·7	N14 47·1$_{45}$	11 55 02·1$_{26}$
Mon.	6	54 28·4	17 00·6$_{41}$	53 59·0$_{15}$		Tues.	6	26 00·8	14 42·6$_{46}$	55 04·7$_{26}$
	12	55 27·5	16 56·5$_{40}$	54 00·5$_{15}$			12	27 00·0	14 38·0$_{46}$	55 07·3$_{27}$
	18	56 26·7	16 52·5$_{41}$	54 02·0$_{16}$			18	27 59·1	14 33·4$_{45}$	55 10·0$_{27}$
6	0	20 57 25·8	N16 48·4$_{41}$	11 54 03·6$_{16}$		14	0	21 28 58·3	N14 28·9$_{47}$	11 55 12·7$_{28}$
Tues.	6	58 25·0	16 44·3$_{42}$	54 05·2$_{16}$		Wed.	6	29 57·4	14 24·2$_{46}$	55 15·5$_{28}$
	12	20 59 24·1	16 40·1$_{41}$	54 06·8$_{17}$			12	30 56·5	14 19·6$_{46}$	55 18·3$_{28}$
	18	21 00 23·2	16 36·0$_{42}$	54 08·5$_{17}$			18	31 55·7	14 15·0$_{47}$	55 21·1$_{29}$
7	0	21 01 22·4	N16 31·8$_{42}$	11 54 10·2$_{17}$		15	0	21 32 54·8	N14 10·3$_{46}$	11 55 24·0$_{29}$
Wed.	6	02 21·5	16 27·6$_{42}$	54 11·9$_{18}$		Thur.	6	33 53·9	14 05·7$_{47}$	55 26·9$_{29}$
	12	03 20·7	16 23·4$_{42}$	54 13·7$_{18}$			12	34 53·1	14 01·0$_{47}$	55 29·8$_{30}$
	18	04 19·8	16 19·2$_{42}$	54 15·5$_{19}$			18	35 52·2	13 56·3$_{47}$	55 32·8$_{30}$
8	0	21 05 18·9	N16 15·0$_{42}$	11 54 17·4$_{19}$		16	0	21 36 51·4	N13 51·6$_{48}$	11 55 35·8$_{30}$
Thur.	6	06 18·1	16 10·8$_{43}$	54 19·3$_{19}$		Fri.	6	37 50·5	13 46·8$_{47}$	55 38·8$_{31}$
	12	07 17·2	16 06·5$_{43}$	54 21·2$_{20}$			12	38 49·6	13 42·1$_{47}$	55 41·9$_{31}$
	18	08 16·4	16 02·2$_{43}$	54 23·2$_{19}$			18	39 48·8	13 37·4$_{48}$	55 45·0$_{31}$
	24	21 09 15·5	N15 57·9$_{43}$	11 54 25·1			24	21 40 47·9	N13 32·6$_{48}$	11 55 48·1$_{31}$

Sun's SD   15·8

SUNRISE

Date	South Latitude								0°	North Latitude							
	60°	55°	50°	45°	40°	30°	20°	10°	0°	10°	20°	30°	40°	45°	50°	55°	60°
Aug.	h	h	h	h	h	h	h	h	h	h	h	h	h	h	h	h	h
1	8·3	7·8	7·5	7·3	7·1	6·8	6·5	6·3	6·0	5·8	5·6	5·3	5·0	4·7	4·5	4·1	3·6
6	8·1	7·7	7·4	7·2	7·0	6·7	6·5	6·2	6·0	5·8	5·6	5·4	5·0	4·8	4·6	4·3	3·9
11	7·8	7·5	7·3	7·1	6·9	6·6	6·4	6·2	6·0	5·8	5·7	5·4	5·1	4·9	4·7	4·4	4·0
16	7·6	7·3	7·1	6·9	6·8	6·5	6·3	6·2	6·0	5·8	5·7	5·5	5·2	5·0	4·8	4·6	4·2
21	7·4	7·2	7·0	6·8	6·7	6·5	6·3	6·2	6·0	5·8	5·7	5·5	5·3	5·1	5·0	4·7	4·5
26	7·2	7·0	6·8	6·7	6·5	6·4	6·2	6·1	6·0	5·8	5·7	5·5	5·3	5·2	5·1	4·9	4·7
31	6·9	6·7	6·6	6·5	6·4	6·3	6·2	6·0	6·0	5·8	5·7	5·6	5·4	5·3	5·2	5·0	4·8

Moon's Phases: Last Quarter 1^d 10^h 22^m New Moon 8^d 19^h 15^m First Quarter 15^d 10^h 12^m

UT		R	Dec	E	UT		R	Dec	E
d	h	h m s	° ′	h m s	d	h	h m s	° ′	h m s
17	0	21 40 47·9	N13 32·6 ₄₈	11 55 48·1 ₃₂	**25**	0	22 12 20·4	N10 53·6 ₅₂	11 57 44·8 ₄₁
Sat.	6	41 47·1	13 27·8 ₄₈	55 51·3 ₃₂	**Sun.**	6	13 19·5	10 48·4 ₅₁	57 48·9 ₄₁
	12	42 46·2	13 23·0 ₄₈	55 54·5 ₃₂		12	14 18·6	10 43·3 ₅₂	57 53·0 ₄₂
	18	43 45·3	13 18·2 ₄₈	55 57·7 ₃₃		18	15 17·8	10 38·1 ₅₂	57 57·2 ₄₂
18	0	21 44 44·5	N13 13·4 ₄₈	11 56 01·0 ₃₃	**26**	0	22 16 16·9	N10 32·9 ₅₂	11 58 01·4 ₄₂
Sun.	6	45 43·6	13 08·6 ₄₉	56 04·3 ₃₃	**Mon.**	6	17 16·1	10 27·7 ₅₂	58 05·6 ₄₂
	12	46 42·8	13 03·7 ₄₈	56 07·6 ₃₄		12	18 15·2	10 22·5 ₅₂	58 09·8 ₄₃
	18	47 41·9	12 58·9 ₄₉	56 11·0 ₃₄		18	19 14·3	10 17·3 ₅₃	58 14·1 ₄₃
19	0	21 48 41·0	N12 54·0 ₄₉	11 56 14·4 ₃₄	**27**	0	22 20 13·5	N10 12·0 ₅₂	11 58 18·4 ₄₃
Mon.	6	49 40·2	12 49·1 ₄₉	56 17·8 ₃₅	**Tues.**	6	21 12·6	10 06·8 ₅₃	58 22·7 ₄₃
	12	50 39·3	12 44·2 ₄₉	56 21·3 ₃₄		12	22 11·7	10 01·5 ₅₂	58 27·0 ₄₄
	18	51 38·5	12 39·3 ₄₉	56 24·7 ₃₆		18	23 10·9	9 56·3 ₅₃	58 31·4 ₄₃
20	0	21 52 37·6	N12 34·4 ₄₉	11 56 28·3 ₃₅	**28**	0	22 24 10·0	N 9 51·0 ₅₃	11 58 35·7 ₄₄
Tues.	6	53 36·7	12 29·5 ₄₉	56 31·8 ₃₆	**Wed.**	6	25 09·2	9 45·7 ₅₃	58 40·1 ₄₅
	12	54 35·9	12 24·6 ₅₀	56 35·4 ₃₆		12	26 08·3	9 40·4 ₅₃	58 44·6 ₄₅
	18	55 35·0	12 19·6 ₅₀	56 39·0 ₃₇		18	27 07·4	9 35·1 ₅₃	58 49·0 ₄₅
21	0	21 56 34·2	N12 14·6 ₄₉	11 56 42·7 ₃₆	**29**	0	22 28 06·6	N 9 29·8 ₅₃	11 58 53·5 ₄₄
Wed.	6	57 33·3	12 09·7 ₅₀	56 46·3 ₃₇	**Thur.**	6	29 05·7	9 24·5 ₅₄	58 57·9 ₄₅
	12	58 32·4	12 04·7 ₅₀	56 50·0 ₃₈		12	30 04·8	9 19·1 ₅₄	59 02·4 ₄₆
	18	21 59 31·6	11 59·7 ₅₀	56 53·8 ₃₇		18	31 04·0	9 13·8 ₅₄	59 07·0 ₄₅
22	0	22 00 30·7	N11 54·7 ₅₁	11 56 57·5 ₃₈	**30**	0	22 32 03·1	N 9 08·4 ₅₃	11 59 11·5 ₄₆
Thur.	6	01 29·9	11 49·6 ₅₀	57 01·3 ₃₈	**Fri.**	6	33 02·3	9 03·1 ₅₄	59 16·1 ₄₆
	12	02 29·0	11 44·6 ₅₁	57 05·1 ₃₉		12	34 01·4	8 57·7 ₅₄	59 20·7 ₄₆
	18	03 28·1	11 39·5 ₅₀	57 09·0 ₃₈		18	35 00·5	8 52·3 ₅₃	59 25·3 ₄₆
23	0	22 04 27·3	N11 34·5 ₅₁	11 57 12·8 ₃₉	**31**	0	22 35 59·7	N 8 47·0 ₅₄	11 59 29·9 ₄₆
Fri.	6	05 26·4	11 29·4 ₅₁	57 16·7 ₄₀	**Sat.**	6	36 58·8	8 41·6 ₅₄	59 34·5 ₄₇
	12	06 25·5	11 24·3 ₅₁	57 20·7 ₃₉		12	37 58·0	8 36·2 ₅₅	59 39·2 ₄₇
	18	07 24·7	11 19·2 ₅₁	57 24·6 ₄₀		18	38 57·1	8 30·7 ₅₄	59 43·9 ₄₇
24	0	22 08 23·8	N11 14·1 ₅₁	11 57 28·6 ₄₀	**32**	0	22 39 56·2	N 8 25·3 ₅₄	11 59 48·6 ₄₇
Sat.	6	09 23·0	11 09·0 ₅₁	57 32·6 ₄₀	**Sun.**	6	40 55·4	8 19·9 ₅₄	59 53·3 ₄₇
	12	10 22·1	11 03·9 ₅₂	57 36·6 ₄₁		12	41 54·5	8 14·5 ₅₅	11 59 58·0 ₄₈
	18	11 21·2	10 58·7 ₅₁	57 40·7 ₄₁		18	42 53·7	8 09·0 ₅₄	12 00 02·8 ₄₇
	24	22 12 20·4	N10 53·6	11 57 44·8		24	22 43 52·8	N 8 03·6	12 00 07·5

Sun's SD   15·8

SUNSET

Date	South Latitude									North Latitude							
	60°	55°	50°	45°	40°	30°	20°	10°	0°	10°	20°	30°	40°	45°	50°	55°	60°
Aug.	h	h	h	h	h	h	h	h	h	h	h	h	h	h	h	h	h
1	16·0	16·4	16·7	16·9	17·1	17·4	17·7	17·9	18·2	18·4	18·6	18·9	19·2	19·4	19·7	20·1	20·5
6	16·1	16·5	16·8	17·0	17·2	17·5	17·7	17·9	18·1	18·4	18·6	18·8	19·1	19·4	19·6	19·9	20·3
11	16·3	16·7	16·9	17·1	17·3	17·6	17·8	18·0	18·1	18·3	18·5	18·7	19·1	19·2	19·4	19·7	20·1
16	16·5	16·8	17·0	17·2	17·4	17·6	17·8	18·0	18·1	18·3	18·5	18·7	18·9	19·1	19·3	19·5	19·9
21	16·7	17·0	17·1	17·3	17·4	17·6	17·8	18·0	18·1	18·2	18·4	18·6	18·8	18·9	19·1	19·3	19·6
26	16·9	17·1	17·3	17·4	17·5	17·7	17·8	18·0	18·1	18·2	18·4	18·5	18·7	18·8	18·9	19·1	19·4
31	17·1	17·3	17·4	17·5	17·6	17·7	17·9	18·0	18·1	18·2	18·3	18·4	18·6	18·7	18·8	18·9	19·1

Moon's Phases: Full Moon 22ᵈ 22ʰ 29ᵐ   Last Quarter 31ᵈ 02ʰ 31ᵐ

UT		R	Dec	E	UT		R	Dec	E
d	h	h m s	° '	h m s	d	h	h m s	° '	h m s
1	0	22 39 56·2	N 8 25·3 ₅₄	11 59 48·6 ₄₇	**9**	0	23 11 28·7	N 5 27·8 ₅₇	12 02 27·1 ₅₁
Sun.	6	40 55·4	8 19·9 ₅₄	59 53·3 ₄₇	Mon.	6	12 27·8	5 22·1 ₅₆	02 32·2 ₅₂
	12	41 54·5	8 14·5 ₅₅	11 59 58·0 ₄₈		12	13 26·9	5 16·5 ₅₇	02 37·4 ₅₂
	18	42 53·7	8 09·0 ₅₄	12 00 02·8 ₄₇		18	14 26·1	5 10·8 ₅₇	02 42·6 ₅₂
2	0	22 43 52·8	N 8 03·6 ₅₅	12 00 07·5 ₄₈	**10**	0	23 15 25·2	N 5 05·1 ₅₇	12 02 47·8 ₅₂
Mon.	6	44 51·9	7 58·1 ₅₅	00 12·3 ₄₈	Tues.	6	16 24·4	4 59·4 ₅₆	02 53·0 ₅₂
	12	45 51·1	7 52·6 ₅₅	00 17·1 ₄₈		12	17 23·5	4 53·8 ₅₇	02 58·2 ₅₃
	18	46 50·2	7 47·1 ₅₅	00 21·9 ₄₉		18	18 22·6	4 48·1 ₅₇	03 03·5 ₅₂
3	0	22 47 49·4	N 7 41·6 ₅₄	12 00 26·8 ₄₈	**11**	0	23 19 21·8	N 4 42·4 ₅₇	12 03 08·7 ₅₃
Tues.	6	48 48·5	7 36·2 ₅₆	00 31·6 ₄₉	Wed.	6	20 20·9	4 36·7 ₅₇	03 14·0 ₅₂
	12	49 47·6	7 30·6 ₅₅	00 36·5 ₄₉		12	21 20·0	4 31·0 ₅₇	03 19·2 ₅₃
	18	50 46·8	7 25·1 ₅₅	00 41·4 ₄₉		18	22 19·2	4 25·3 ₅₈	03 24·5 ₅₂
4	0	22 51 45·9	N 7 19·6 ₅₅	12 00 46·3 ₄₉	**12**	0	23 23 18·3	N 4 19·5 ₅₇	12 03 29·7 ₅₃
Wed.	6	52 45·1	7 14·1 ₅₅	00 51·2 ₄₉	Thur.	6	24 17·5	4 13·8 ₅₇	03 35·0 ₅₃
	12	53 44·2	7 08·6 ₅₆	00 56·1 ₄₉		12	25 16·6	4 08·1 ₅₇	03 40·3 ₅₃
	18	54 43·3	7 03·0 ₅₅	01 01·0 ₅₀		18	26 15·7	4 02·4 ₅₈	03 45·6 ₅₃
5	0	22 55 42·5	N 6 57·5 ₅₆	12 01 06·0 ₅₀	**13**	0	23 27 14·9	N 3 56·6 ₅₇	12 03 50·9 ₅₃
Thur.	6	56 41·6	6 51·9 ₅₅	01 11·0 ₅₀	Fri.	6	28 14·0	3 50·9 ₅₇	03 56·2 ₅₃
	12	57 40·7	6 46·4 ₅₆	01 16·0 ₅₀		12	29 13·2	3 45·2 ₅₈	04 01·5 ₅₃
	18	58 39·9	6 40·8 ₅₆	01 21·0 ₅₀		18	30 12·3	3 39·4 ₅₇	04 06·8 ₅₄
6	0	22 59 39·0	N 6 35·2 ₅₆	12 01 26·0 ₅₀	**14**	0	23 31 11·4	N 3 33·7 ₅₈	12 04 12·2 ₅₃
Fri.	6	23 00 38·2	6 29·6 ₅₆	01 31·0 ₅₀	Sat.	6	32 10·6	3 27·9 ₅₇	04 17·5 ₅₃
	12	01 37·3	6 24·0 ₅₆	01 36·0 ₅₁		12	33 09·7	3 22·2 ₅₈	04 22·8 ₅₄
	18	02 36·4	6 18·4 ₅₆	01 41·1 ₅₀		18	34 08·9	3 16·4 ₅₈	04 28·2 ₅₃
7	0	23 03 35·6	N 6 12·8 ₅₆	12 01 46·1 ₅₁	**15**	0	23 35 08·0	N 3 10·6 ₅₇	12 04 33·5 ₅₃
Sat.	6	04 34·7	6 07·2 ₅₆	01 51·2 ₅₁	Sun.	6	36 07·1	3 04·9 ₅₈	04 38·8 ₅₄
	12	05 33·8	6 01·6 ₅₆	01 56·3 ₅₁		12	37 06·3	2 59·1 ₅₈	04 44·2 ₅₃
	18	06 33·0	5 56·0 ₅₆	02 01·4 ₅₁		18	38 05·4	2 53·3 ₅₈	04 49·5 ₅₄
8	0	23 07 32·1	N 5 50·4 ₅₇	12 02 06·5 ₅₁	**16**	0	23 39 04·6	N 2 47·5 ₅₇	12 04 54·9 ₅₄
Sun.	6	08 31·3	5 44·7 ₅₆	02 11·6 ₅₂	Mon.	6	40 03·7	2 41·8 ₅₈	05 00·3 ₅₃
	12	09 30·4	5 39·1 ₅₇	02 16·8 ₅₁		12	41 02·8	2 36·0 ₅₈	05 05·6 ₅₄
	18	10 29·5	5 33·4 ₅₇	02 21·9 ₅₂		18	42 02·0	2 30·2 ₅₈	05 11·0 ₅₄
	24	23 11 28·7	N 5 27·8 ₅₆	12 02 27·1		24	23 43 01·1	N 2 24·4 ₅₈	12 05 16·3 ₅₃

Sun's SD   15·9

SUNRISE

Date	South Latitude								0°	North Latitude							
	60°	55°	50°	45°	40°	30°	20°	10°		10°	20°	30°	40°	45°	50°	55°	60°
Sept.	h	h	h	h	h	h	h	h	h	h	h	h	h	h	h	h	h
1	6·9	6·7	6·6	6·5	6·4	6·3	6·2	6·0	6·0	5·8	5·7	5·6	5·5	5·3	5·2	5·1	4·9
6	6·6	6·5	6·4	6·3	6·3	6·2	6·1	6·0	5·9	5·8	5·7	5·7	5·5	5·5	5·4	5·2	5·1
11	6·4	6·3	6·2	6·2	6·1	6·1	6·0	6·0	5·9	5·8	5·8	5·7	5·6	5·5	5·5	5·4	5·3
16	6·1	6·1	6·0	6·0	6·0	6·0	5·9	5·9	5·9	5·8	5·8	5·7	5·7	5·7	5·6	5·5	5·5
21	5·9	5·9	5·9	5·9	5·9	5·8	5·8	5·8	5·8	5·8	5·8	5·8	5·8	5·7	5·7	5·7	5·7
26	5·6	5·7	5·7	5·7	5·7	5·7	5·8	5·8	5·8	5·8	5·8	5·8	5·8	5·8	5·9	5·9	5·9
31	5·4	5·5	5·5	5·5	5·6	5·7	5·7	5·7	5·8	5·8	5·8	5·9	5·9	6·0	6·0	6·0	6·1

Moon's Phases: New Moon 7d 03h 10m   First Quarter 13d 18h 08m

UT d h	R h m s	Dec ° '	E h m s
17 0	23 43 01·1	N 2 24·4	12 05 16·3
Tues. 6	44 00·3	2 18·6 ₅₈	05 21·7 ₅₄
12	44 59·4	2 12·8 ₅₈	05 27·1 ₅₄
18	45 58·5	2 07·0 ₅₈	05 32·4 ₅₃ ₅₄
18 0	23 46 57·7	N 2 01·2	12 05 37·8
Wed. 6	47 56·8	1 55·4 ₅₈	05 43·2 ₅₄
12	48 55·9	1 49·6 ₅₈	05 48·5 ₅₃
18	49 55·1	1 43·8 ₅₈	05 53·9 ₅₄
19 0	23 50 54·2	N 1 38·0	12 05 59·3
Thur. 6	51 53·4	1 32·2 ₅₈	06 04·6 ₅₃ ₅₄
12	52 52·5	1 26·4 ₅₈	06 10·0 ₅₃
18	53 51·6	1 20·5 ₅₉ ₅₈	06 15·3 ₅₄
20 0	23 54 50·8	N 1 14·7	12 06 20·7
Fri. 6	55 49·9	1 08·9 ₅₈	06 26·1 ₅₄
12	56 49·0	1 03·1 ₅₈	06 31·4 ₅₃
18	57 48·2	0 57·3 ₅₈ ₅₉	06 36·7 ₅₃ ₅₄
21 0	23 58 47·3	N 0 51·4	12 06 42·1
Sat. 6	23 59 46·5	0 45·6 ₅₈	06 47·4 ₅₃
12	0 00 45·6	0 39·8 ₅₈	06 52·7 ₅₃
18	01 44·7	0 33·9 ₅₉ ₅₈	06 58·1 ₅₄ ₅₃
22 0	0 02 43·9	N 0 28·1	12 07 03·4
Sun. 6	03 43·0	0 22·3 ₅₈	07 08·7 ₅₃
12	04 42·1	0 16·5 ₅₈	07 14·0 ₅₃
18	05 41·3	0 10·6 ₅₉ ₅₈	07 19·3 ₅₃ ₅₃
23 0	0 06 40·4	N 0 04·8	12 07 24·6
Mon. 6	07 39·5	S 0 01·1 ₅₉	07 29·9 ₅₃
12	08 38·7	0 06·9 ₅₈	07 35·2 ₅₃
18	09 37·8	0 12·7 ₅₈ ₅₉	07 40·4 ₅₂ ₅₃
24 0	0 10 37·0	S 0 18·6	12 07 45·7
Tues. 6	11 36·1	0 24·4 ₅₈ ₅₉	07 50·9 ₅₂
12	12 35·2	0 30·3 ₅₈	07 56·2 ₅₃
18	13 34·4	0 36·1 ₅₈	08 01·4 ₅₂
24	0 14 33·5	S 0 41·9 ₅₈	12 08 06·6 ₅₂

UT d h	R h m s	Dec ° '	E h m s
25 0	0 14 33·5	S 0 41·9	12 08 06·6
Wed. 6	15 32·6	0 47·8 ₅₉	08 11·9 ₅₃
12	16 31·8	0 53·6 ₅₈	08 17·1 ₅₂
18	17 30·9	0 59·5 ₅₉ ₅₈	08 22·2 ₅₁ ₅₂
26 0	0 18 30·1	S 1 05·3	12 08 27·4
Thur. 6	19 29·2	1 11·1 ₅₈	08 32·6 ₅₂
12	20 28·3	1 17·0 ₅₉	08 37·7 ₅₁
18	21 27·5	1 22·8 ₅₈ ₅₉	08 42·9 ₅₂ ₅₁
27 0	0 22 26·6	S 1 28·7	12 08 48·0
Fri. 6	23 25·7	1 34·5 ₅₈	08 53·1 ₅₁
12	24 24·9	1 40·3 ₅₈	08 58·2 ₅₁
18	25 24·0	1 46·2 ₅₉ ₅₈	09 03·3 ₅₁ ₅₁
28 0	0 26 23·2	S 1 52·0	12 09 08·4
Sat. 6	27 22·3	1 57·9 ₅₉	09 13·5 ₅₁
12	28 21·4	2 03·7 ₅₈	09 18·5 ₅₀
18	29 20·6	2 09·5 ₅₈ ₅₉	09 23·5 ₅₀ ₅₀
29 0	0 30 19·7	S 2 15·4	12 09 28·5
Sun. 6	31 18·9	2 21·2 ₅₈	09 33·5 ₅₀
12	32 18·0	2 27·0 ₅₈	09 38·5 ₅₀
18	33 17·1	2 32·9 ₅₉ ₅₈	09 43·5 ₅₀ ₄₉
30 0	0 34 16·3	S 2 38·7	12 09 48·4
Mon. 6	35 15·4	2 44·5 ₅₈	09 53·4 ₅₀
12	36 14·6	2 50·3 ₅₈	09 58·3 ₄₉
18	37 13·7	2 56·2 ₅₉ ₅₈	10 03·2 ₄₉ ₄₈
31 0	0 38 12·8	S 3 02·0	12 10 08·0
Tues. 6	39 12·0	3 07·8 ₅₈	10 12·9 ₄₉
12	40 11·1	3 13·6 ₅₈	10 17·7 ₄₈
18	41 10·3	3 19·4 ₅₈ ₅₉	10 22·6 ₄₉ ₄₈
24	0 42 09·4	S 3 25·3	12 10 27·4

Sun's SD   16'·0

SUNSET

Date	South Latitude									North Latitude							
	60°	55°	50°	45°	40°	30°	20°	10°	0°	10°	20°	30°	40°	45°	50°	55°	60°
Sept.	h	h	h	h	h	h	h	h	h	h	h	h	h	h	h	h	h
1	17·1	17·3	17·4	17·5	17·6	17·7	17·9	18·0	18·1	18·1	18·3	18·4	18·5	18·6	18·7	18·9	19·1
6	17·3	17·4	17·6	17·6	17·7	17·8	17·9	17·9	18·0	18·1	18·2	18·3	18·4	18·5	18·6	18·7	18·8
11	17·5	17·6	17·7	17·7	17·8	17·8	17·9	17·9	18·0	18·1	18·1	18·2	18·3	18·3	18·4	18·5	18·6
16	17·7	17·8	17·8	17·8	17·9	17·9	17·9	17·9	18·0	18·0	18·0	18·1	18·1	18·2	18·2	18·2	18·3
21	17·9	17·9	17·9	17·9	17·9	17·9	17·9	17·9	17·9	17·9	18·0	18·0	18·0	18·0	18·0	18·1	18·1
26	18·1	18·1	18·1	18·0	18·0	18·0	17·9	17·9	17·9	17·9	17·9	17·9	17·9	17·9	17·8	17·8	17·8
31	18·3	18·2	18·2	18·1	18·1	18·0	18·0	17·9	17·9	17·9	17·8	17·8	17·7	17·7	17·6	17·6	17·6

Moon's Phases: Full Moon 21d 13h 59m   Last Quarter 29d 17h 03m

SUN – OCTOBER, 2002

UT d h	R h m s	Dec ° '	E h m s
1 0	0 38 12·8 $_{58}$	S 3 02·0 $_{58}$	12 10 08·0 $_{49}$
Tues. 6	39 12·0 $_{58}$	3 07·8 $_{58}$	10 12·9 $_{48}$
12	40 11·1 $_{58}$	3 13·6 $_{58}$	10 17·7 $_{49}$
18	41 10·3 $_{59}$	3 19·4 $_{58}$	10 22·6 $_{48}$
2 0	0 42 09·4 $_{58}$	S 3 25·3 $_{58}$	12 10 27·4 $_{47}$
Wed. 6	43 08·5 $_{58}$	3 31·1 $_{58}$	10 32·1 $_{48}$
12	44 07·7 $_{58}$	3 36·9 $_{58}$	10 36·9 $_{47}$
18	45 06·8 $_{58}$	3 42·7 $_{58}$	10 41·6 $_{48}$
3 0	0 46 06·0 $_{58}$	S 3 48·5 $_{58}$	12 10 46·4 $_{47}$
Thur. 6	47 05·1 $_{58}$	3 54·3 $_{58}$	10 51·1 $_{46}$
12	48 04·2 $_{58}$	4 00·1 $_{58}$	10 55·7 $_{47}$
18	49 03·4 $_{58}$	4 05·9 $_{58}$	11 00·4 $_{46}$
4 0	0 50 02·5 $_{58}$	S 4 11·7 $_{58}$	12 11 05·0 $_{46}$
Fri. 6	51 01·6 $_{58}$	4 17·5 $_{58}$	11 09·6 $_{46}$
12	52 00·8 $_{57}$	4 23·3 $_{57}$	11 14·2 $_{46}$
18	52 59·9 $_{58}$	4 29·0 $_{58}$	11 18·8 $_{45}$
5 0	0 53 59·1 $_{58}$	S 4 34·8 $_{58}$	12 11 23·3 $_{46}$
Sat. 6	54 58·2 $_{58}$	4 40·6 $_{58}$	11 27·9 $_{45}$
12	55 57·3 $_{58}$	4 46·4 $_{58}$	11 32·4 $_{44}$
18	56 56·5 $_{57}$	4 52·2 $_{57}$	11 36·8 $_{45}$
6 0	0 57 55·6 $_{58}$	S 4 57·9 $_{58}$	12 11 41·3 $_{44}$
Sun. 6	58 54·7 $_{57}$	5 03·7 $_{57}$	11 45·7 $_{44}$
12	0 59 53·9 $_{58}$	5 09·4 $_{58}$	11 50·1 $_{44}$
18	1 00 53·0 $_{58}$	5 15·2 $_{58}$	11 54·5 $_{44}$
7 0	1 01 52·1 $_{57}$	S 5 21·0 $_{57}$	12 11 58·9 $_{43}$
Mon. 6	02 51·3 $_{57}$	5 26·7 $_{57}$	12 03·2 $_{43}$
12	03 50·4 $_{58}$	5 32·4 $_{58}$	12 07·5 $_{43}$
18	04 49·6 $_{57}$	5 38·2 $_{57}$	12 11·8 $_{42}$
8 0	1 05 48·7 $_{57}$	S 5 43·9 $_{57}$	12 12 16·0 $_{42}$
Tues. 6	06 47·8 $_{58}$	5 49·6 $_{58}$	12 20·2 $_{43}$
12	07 47·0 $_{57}$	5 55·4 $_{57}$	12 24·5 $_{41}$
18	08 46·1 $_{57}$	6 01·1 $_{57}$	12 28·6 $_{42}$
24	1 09 45·2	S 6 06·8	12 12 32·8

UT d h	R h m s	Dec ° '	E h m s
9 0	1 09 45·2 $_{57}$	S 6 06·8 $_{57}$	12 12 32·8 $_{41}$
Wed. 6	10 44·4 $_{57}$	6 12·5 $_{57}$	12 36·9 $_{41}$
12	11 43·5 $_{57}$	6 18·2 $_{57}$	12 41·0 $_{41}$
18	12 42·7 $_{57}$	6 23·9 $_{57}$	12 45·1 $_{40}$
10 0	1 13 41·8 $_{57}$	S 6 29·6 $_{57}$	12 12 49·1 $_{40}$
Thur. 6	14 40·9 $_{57}$	6 35·3 $_{57}$	12 53·1 $_{40}$
12	15 40·1 $_{56}$	6 41·0 $_{57}$	12 57·1 $_{40}$
18	16 39·2 $_{57}$	6 46·7 $_{56}$	13 01·1 $_{39}$
11 0	1 17 38·4 $_{57}$	S 6 52·3 $_{57}$	12 13 05·0 $_{39}$
Fri. 6	18 37·5 $_{56}$	6 58·0 $_{56}$	13 08·9 $_{39}$
12	19 36·6 $_{57}$	7 03·6 $_{57}$	13 12·8 $_{38}$
18	20 35·8 $_{56}$	7 09·3 $_{56}$	13 16·6 $_{38}$
12 0	1 21 34·9 $_{57}$	S 7 14·9 $_{57}$	12 13 20·4 $_{38}$
Sat. 6	22 34·1 $_{56}$	7 20·6 $_{56}$	13 24·2 $_{38}$
12	23 33·2 $_{57}$	7 26·2 $_{57}$	13 28·0 $_{37}$
18	24 32·3 $_{56}$	7 31·9 $_{56}$	13 31·7 $_{37}$
13 0	1 25 31·5 $_{56}$	S 7 37·5 $_{56}$	12 13 35·4 $_{36}$
Sun. 6	26 30·6 $_{56}$	7 43·1 $_{56}$	13 39·0 $_{37}$
12	27 29·8 $_{56}$	7 48·7 $_{56}$	13 42·7 $_{36}$
18	28 28·9 $_{56}$	7 54·3 $_{56}$	13 46·3 $_{35}$
14 0	1 29 28·0 $_{56}$	S 7 59·9 $_{56}$	12 13 49·8 $_{36}$
Mon. 6	30 27·2 $_{56}$	8 05·5 $_{56}$	13 53·4 $_{35}$
12	31 26·3 $_{55}$	8 11·1 $_{55}$	13 56·9 $_{34}$
18	32 25·5 $_{56}$	8 16·6 $_{56}$	14 00·3 $_{35}$
15 0	1 33 24·6 $_{56}$	S 8 22·2 $_{56}$	12 14 03·8 $_{34}$
Tues. 6	34 23·7 $_{55}$	8 27·8 $_{55}$	14 07·2 $_{33}$
12	35 22·9 $_{55}$	8 33·3 $_{55}$	14 10·5 $_{34}$
18	36 22·0 $_{56}$	8 38·8 $_{56}$	14 13·9 $_{33}$
16 0	1 37 21·2 $_{55}$	S 8 44·4 $_{55}$	12 14 17·2 $_{32}$
Wed. 6	38 20·3 $_{55}$	8 49·9 $_{55}$	14 20·4 $_{32}$
12	39 19·4 $_{55}$	8 55·4 $_{55}$	14 23·6 $_{32}$
18	40 18·6 $_{55}$	9 00·9 $_{55}$	14 26·8 $_{32}$
24	1 41 17·7	S 9 06·4	12 14 30·0

Sun's SD 16'·0

SUNRISE

Date	South Latitude									North Latitude							
	60°	55°	50°	45°	40°	30°	20°	10°	0°	10°	20°	30°	40°	45°	50°	55°	60°
Oct.	h	h	h	h	h	h	h	h	h	h	h	h	h	h	h	h	h
1	5·4	5·5	5·5	5·5	5·6	5·7	5·7	5·7	5·8	5·8	5·8	5·9	5·9	6·0	6·0	6·0	6·1
6	5·1	5·2	5·3	5·4	5·5	5·5	5·6	5·7	5·7	5·8	5·9	5·9	6·0	6·1	6·1	6·2	6·3
11	4·9	5·0	5·2	5·2	5·3	5·5	5·5	5·7	5·7	5·8	5·9	6·0	6·1	6·2	6·2	6·3	6·5
16	4·6	4·8	5·0	5·1	5·2	5·3	5·5	5·6	5·7	5·8	5·9	6·0	6·2	6·3	6·4	6·5	6·7
21	4·4	4·6	4·8	5·0	5·1	5·3	5·4	5·6	5·7	5·8	6·0	6·1	6·3	6·4	6·5	6·7	6·9
26	4·1	4·4	4·6	4·8	5·0	5·2	5·4	5·5	5·7	5·8	6·0	6·2	6·4	6·5	6·7	6·8	7·1
31	3·9	4·2	4·5	4·7	4·8	5·1	5·3	5·5	5·7	5·8	6·0	6·2	6·5	6·6	6·8	7·0	7·3

Moon's Phases: New Moon 6^d 11^h 18^m First Quarter 13^d 05^h 33^m

UT d h	R h m s	Dec ° '	E h m s	UT d h	R h m s	Dec ° '	E h m s
17 0	1 41 17·7	S 9 06·4 (55)	12 14 30·0 (31)	**25** 0	2 12 50·1	S 11 57·6 (52)	12 15 50·4 (19)
Thur. 6	42 16·8	9 11·9 (55)	14 33·1 (31)	**Fri.** 6	13 49·2	12 02·8 (52)	15 52·3 (17)
12	43 16·0	9 17·4 (55)	14 36·2 (31)	12	14 48·4	12 08·0 (51)	15 54·0 (18)
18	44 15·1	9 22·9 (55)	14 39·3 (30)	18	15 47·5	12 13·1 (52)	15 55·8 (16)
18 0	1 45 14·3	S 9 28·4 (54)	12 14 42·3 (29)	**26** 0	2 16 46·7	S 12 18·3 (51)	12 15 57·4 (17)
Fri. 6	46 13·4	9 33·8 (55)	14 45·2 (30)	**Sat.** 6	17 45·8	12 23·4 (51)	15 59·1 (15)
12	47 12·5	9 39·3 (54)	14 48·2 (29)	12	18 44·9	12 28·5 (52)	16 00·6 (16)
18	48 11·7	9 44·7 (55)	14 51·1 (28)	18	19 44·1	12 33·7 (51)	16 02·2 (15)
19 0	1 49 10·8	S 9 50·2 (54)	12 14 53·9 (29)	**27** 0	2 20 43·2	S 12 38·8 (50)	12 16 03·7 (14)
Sat. 6	50 09·9	9 55·6 (54)	14 56·8 (27)	**Sun.** 6	21 42·4	12 43·8 (51)	16 05·1 (14)
12	51 09·1	10 01·0 (54)	14 59·5 (28)	12	22 41·5	12 48·9 (51)	16 06·5 (14)
18	52 08·2	10 06·4 (54)	15 02·3 (27)	18	23 40·7	12 54·0 (50)	16 07·9 (13)
20 0	1 53 07·3	S 10 11·8 (54)	12 15 05·0 (27)	**28** 0	2 24 39·8	S 12 59·0 (51)	12 16 09·2 (12)
Sun. 6	54 06·5	10 17·2 (54)	15 07·7 (26)	**Mon.** 6	25 38·9	13 04·1 (50)	16 10·4 (12)
12	55 05·6	10 22·6 (53)	15 10·3 (26)	12	26 38·1	13 09·1 (50)	16 11·6 (12)
18	56 04·8	10 27·9 (54)	15 12·9 (25)	18	27 37·2	13 14·1 (50)	16 12·8 (11)
21 0	1 57 03·9	S 10 33·3 (53)	12 15 15·4 (25)	**29** 0	2 28 36·4	S 13 19·1 (50)	12 16 13·9 (11)
Mon. 6	58 03·0	10 38·6 (54)	15 17·9 (25)	**Tues.** 6	29 35·5	13 24·1 (50)	16 15·0 (10)
12	1 59 02·2	10 44·0 (53)	15 20·4 (24)	12	30 34·6	13 29·1 (49)	16 16·0 (9)
18	2 00 01·3	10 49·3 (53)	15 22·8 (24)	18	31 33·8	13 34·0 (50)	16 16·9 (10)
22 0	2 01 00·4	S 10 54·6 (53)	12 15 25·2 (23)	**30** 0	2 32 32·9	S 13 39·0 (49)	12 16 17·9 (8)
Tues. 6	01 59·6	10 59·9 (53)	15 27·5 (23)	**Wed.** 6	33 32·0	13 43·9 (49)	16 18·7 (8)
12	02 58·7	11 05·2 (53)	15 29·8 (23)	12	34 31·2	13 48·8 (49)	16 19·5 (8)
18	03 57·9	11 10·5 (53)	15 32·1 (22)	18	35 30·3	13 53·7 (49)	16 20·3 (7)
23 0	2 04 57·0	S 11 15·8 (53)	12 15 34·3 (22)	**31** 0	2 36 29·5	S 13 58·6 (49)	12 16 21·0 (7)
Wed. 6	05 56·1	11 21·1 (52)	15 36·5 (21)	**Thur.** 6	37 28·6	14 03·5 (48)	16 21·7 (6)
12	06 55·3	11 26·3 (53)	15 38·6 (21)	12	38 27·7	14 08·3 (49)	16 22·3 (5)
18	07 54·4	11 31·6 (52)	15 40·7 (20)	18	39 26·9	14 13·2 (48)	16 22·8 (6)
24 0	2 08 53·6	S 11 36·8 (52)	12 15 42·7 (20)	**32** 0	2 40 26·0	S 14 18·0 (49)	12 16 23·4 (4)
Thur. 6	09 52·7	11 42·0 (52)	15 44·7 (20)	**Fri.** 6	41 25·2	14 22·9 (48)	16 23·8 (4)
12	10 51·8	11 47·2 (53)	15 46·7 (19)	12	42 24·3	14 27·7 (48)	16 24·2 (4)
18	11 51·0	11 52·5 (51)	15 48·6 (18)	18	43 23·4	14 32·5 (47)	16 24·6 (4)
24	2 12 50·1	S 11 57·6	12 15 50·4	24	2 44 22·6	S 14 37·2	12 16 24·9 (3)

Sun's SD   16·1

SUNSET

Date	South Latitude								0°	North Latitude							
	60°	55°	50°	45°	40°	30°	20°	10°	0°	10°	20°	30°	40°	45°	50°	55°	60°
Oct.	h	h	h	h	h	h	h	h	h	h	h	h	h	h	h	h	h
1	18·3	18·2	18·2	18·1	18·1	18·0	18·0	17·9	17·9	17·9	17·8	17·8	17·7	17·7	17·6	17·6	17·6
6	18·5	18·4	18·3	18·2	18·2	18·1	18·0	17·9	17·9	17·8	17·7	17·7	17·6	17·5	17·5	17·4	17·3
11	18·7	18·6	18·4	18·3	18·2	18·1	18·0	17·9	17·8	17·7	17·7	17·6	17·4	17·4	17·3	17·2	17·1
16	18·9	18·7	18·6	18·4	18·3	18·2	18·0	17·9	17·8	17·7	17·6	17·5	17·3	17·2	17·1	17·0	16·8
21	19·1	18·9	18·7	18·6	18·4	18·2	18·1	17·9	17·8	17·7	17·5	17·4	17·2	17·1	17·0	16·8	16·6
26	19·4	19·1	18·9	18·7	18·5	18·3	18·1	17·9	17·8	17·6	17·5	17·3	17·1	17·0	16·8	16·6	16·4
31	19·6	19·2	19·0	18·8	18·6	18·4	18·1	17·9	17·8	17·6	17·4	17·2	17·0	16·8	16·6	16·4	16·1

Moon's Phases: Full Moon 21^d 07^h 20^m Last Quarter 29^d 05^h 28^m

SUN – NOVEMBER, 2002

UT		R	Dec	E	UT		R	Dec	E
d	h	h m s	° ′	h m s	d	h	h m s	° ′	h m s
1	0	2 40 26·0	S 14 18·0 49	12 16 23·4 4	**9**	0	3 11 58·5	S 16 44·7 43	12 16 12·6 11
Fri.	6	41 25·2	14 22·9 48	16 23·8 4	Sat.	6	12 57·6	16 49·0 43	16 11·5 13
	12	42 24·3	14 27·7 48	16 24·2 4		12	13 56·7	16 53·3 42	16 10·2 13
	18	43 23·4	14 32·5 47	16 24·6 3		18	14 55·9	16 57·5 43	16 08·9 13
2	0	2 44 22·6	S 14 37·2 48	12 16 24·9 3	**10**	0	3 15 55·0	S 17 01·8 42	12 16 07·6 14
Sat.	6	45 21·7	14 42·0 47	16 25·2 2	Sun.	6	16 54·2	17 06·0 42	16 06·2 15
	12	46 20·8	14 46·7 48	16 25·4 1		12	17 53·3	17 10·2 42	16 04·7 15
	18	47 20·0	14 51·5 47	16 25·5 1		18	18 52·4	17 14·4 42	16 03·2 15
3	0	2 48 19·1	S 14 56·2 47	12 16 25·6 1	**11**	0	3 19 51·6	S 17 18·6 42	12 16 01·7 16
Sun.	6	49 18·2	15 00·9 47	16 25·7 0	Mon.	6	20 50·7	17 22·8 41	16 00·1 17
	12	50 17·4	15 05·6 47	16 25·7 1		12	21 49·9	17 26·9 41	15 58·4 17
	18	51 16·5	15 10·3 46	16 25·6 1		18	22 49·0	17 31·0 41	15 56·7 17
4	0	2 52 15·7	S 15 14·9 47	12 16 25·5 1	**12**	0	3 23 48·1	S 17 35·1 41	12 15 55·0 18
Mon.	6	53 14·8	15 19·6 46	16 25·4 2	Tues.	6	24 47·3	17 39·2 41	15 53·2 19
	12	54 13·9	15 24·2 46	16 25·2 2		12	25 46·4	17 43·3 40	15 51·3 19
	18	55 13·1	15 28·8 46	16 24·9 3		18	26 45·6	17 47·3 41	15 49·4 20
5	0	2 56 12·2	S 15 33·4 46	12 16 24·6 4	**13**	0	3 27 44·7	S 17 51·4 40	12 15 47·4 20
Tues.	6	57 11·4	15 38·0 45	16 24·2 4	Wed.	6	28 43·8	17 55·4 40	15 45·4 21
	12	58 10·5	15 42·5 46	16 23·8 4		12	29 43·0	17 59·4 39	15 43·3 21
	18	2 59 09·6	15 47·1 45	16 23·4 5		18	30 42·1	18 03·3 40	15 41·2 22
6	0	3 00 08·8	S 15 51·6 45	12 16 22·9 6	**14**	0	3 31 41·2	S 18 07·3 39	12 15 39·0 22
Wed.	6	01 07·9	15 56·1 45	16 22·3 6	Thur.	6	32 40·4	18 11·2 39	15 36·8 23
	12	02 07·1	16 00·6 45	16 21·7 7		12	33 39·5	18 15·1 39	15 34·5 23
	18	03 06·2	16 05·1 45	16 21·0 7		18	34 38·7	18 19·0 39	15 32·2 24
7	0	3 04 05·3	S 16 09·6 44	12 16 20·3 8	**15**	0	3 35 37·8	S 18 22·9 38	12 15 29·8 24
Thur.	6	05 04·5	16 14·0 44	16 19·5 8	Fri.	6	36 36·9	18 26·7 39	15 27·4 25
	12	06 03·6	16 18·4 45	16 18·7 9		12	37 36·1	18 30·6 38	15 24·9 25
	18	07 02·8	16 22·9 44	16 17·8 9		18	38 35·2	18 34·4 38	15 22·4 26
8	0	3 08 01·9	S 16 27·3 43	12 16 16·9 10	**16**	0	3 39 34·3	S 18 38·2 37	12 15 19·8 27
Fri.	6	09 01·0	16 31·6 44	16 15·9 10	Sat.	6	40 33·5	18 41·9 38	15 17·1 27
	12	10 00·2	16 36·0 43	16 14·9 11		12	41 32·6	18 45·7 37	15 14·4 27
	18	10 59·3	16 40·3 44	16 13·8 12		18	42 31·8	18 49·4 37	15 11·7 28
	24	3 11 58·5	S 16 44·7	12 16 12·6		24	3 43 30·9	S 18 53·1	12 15 08·9

Sun's SD   16ʹ2

SUNRISE

Date	South Latitude									North Latitude							
	60°	55°	50°	45°	40°	30°	20°	10°	0°	10°	20°	30°	40°	45°	50°	55°	60°
Nov.	h	h	h	h	h	h	h	h	h	h	h	h	h	h	h	h	h
1	3·9	4·2	4·5	4·7	4·8	5·1	5·3	5·5	5·7	5·8	6·0	6·2	6·5	6·6	6·8	7·0	7·3
6	3·6	4·0	4·3	4·5	4·7	5·0	5·3	5·5	5·7	5·9	6·1	6·3	6·6	6·7	7·0	7·2	7·6
11	3·4	3·9	4·2	4·4	4·7	5·0	5·2	5·5	5·7	5·9	6·1	6·3	6·7	6·9	7·1	7·4	7·8
16	3·2	3·7	4·1	4·3	4·6	4·9	5·2	5·5	5·7	5·9	6·2	6·4	6·8	7·0	7·2	7·5	8·0
21	3·0	3·6	4·0	4·3	4·5	4·9	5·2	5·5	5·7	6·0	6·2	6·5	6·8	7·1	7·4	7·7	8·2
26	2·9	3·5	3·9	4·2	4·5	4·9	5·2	5·5	5·7	6·0	6·2	6·6	7·0	7·2	7·5	7·9	8·4
31	2·7	3·4	3·8	4·2	4·4	4·8	5·2	5·5	5·7	6·0	6·3	6·6	7·0	7·3	7·6	8·0	8·6

Moon's Phases: New Moon 4^d 20^h 34^m First Quarter 11^d 20^h 52^m

UT d	h	R (h m s)	Dec (° ′)	E (h m s)
17	0	3 43 30·9	S18 53·1 ₃₇	12 15 08·9 ₂₉
Sun.	6	44 30·0	18 56·8 ₃₆	15 06·0 ₂₉
	12	45 29·2	19 00·4 ₃₇	15 03·1 ₂₉
	18	46 28·3	19 04·1 ₃₆	15 00·2 ₃₀
18	0	3 47 27·4	S19 07·7 ₃₆	12 14 57·2 ₃₁
Mon.	6	48 26·6	19 11·3 ₃₆	14 54·1 ₃₁
	12	49 25·7	19 14·9 ₃₅	14 51·0 ₃₁
	18	50 24·9	19 18·4 ₃₆	14 47·9 ₃₂
19	0	3 51 24·0	S19 22·0 ₃₅	12 14 44·7 ₃₃
Tues.	6	52 23·1	19 25·5 ₃₅	14 41·4 ₃₃
	12	53 22·3	19 29·0 ₃₄	14 38·1 ₃₃
	18	54 21·4	19 32·4 ₃₅	14 34·7 ₃₄
20	0	3 55 20·6	S19 35·9 ₃₄	12 14 31·3 ₃₅
Wed.	6	56 19·7	19 39·3 ₃₄	14 27·8 ₃₅
	12	57 18·8	19 42·7 ₃₄	14 24·3 ₃₅
	18	58 18·0	19 46·1 ₃₃	14 20·8 ₃₇
21	0	3 59 17·1	S19 49·4 ₃₄	12 14 17·1 ₃₆
Thur.	6	4 00 16·3	19 52·8 ₃₃	14 13·5 ₃₇
	12	01 15·4	19 56·1 ₃₃	14 09·8 ₃₈
	18	02 14·5	19 59·4 ₃₂	14 06·0 ₃₈
22	0	4 03 13·7	S20 02·6 ₃₃	12 14 02·2 ₃₉
Fri.	6	04 12·8	20 05·9 ₃₂	13 58·3 ₃₉
	12	05 12·0	20 09·1 ₃₂	13 54·4 ₄₀
	18	06 11·1	20 12·3 ₃₂	13 50·4 ₄₀
23	0	4 07 10·2	S20 15·5 ₃₁	12 13 46·4 ₄₁
Sat.	6	08 09·4	20 18·6 ₃₁	13 42·3 ₄₁
	12	09 08·5	20 21·7 ₃₁	13 38·2 ₄₂
	18	10 07·7	20 24·8 ₃₁	13 34·0 ₄₂
24	0	4 11 06·8	S20 27·9 ₃₁	12 13 29·8 ₄₃
Sun.	6	12 05·9	20 31·0 ₃₀	13 25·5 ₄₃
	12	13 05·1	20 34·0 ₃₀	13 21·2 ₄₃
	18	14 04·2	20 37·0 ₃₀	13 16·9 ₄₅
	24	4 15 03·4	S20 40·0	12 13 12·4

UT d	h	R (h m s)	Dec (° ′)	E (h m s)
25	0	4 15 03·4	S20 40·0 ₃₀	12 13 12·4 ₄₄
Mon.	6	16 02·5	20 43·0 ₂₉	13 08·0 ₄₅
	12	17 01·6	20 45·9 ₂₉	13 03·5 ₄₆
	18	18 00·8	20 48·8 ₂₉	12 58·9 ₄₆
26	0	4 18 59·9	S20 51·7 ₂₉	12 12 54·3 ₄₇
Tues.	6	19 59·1	20 54·6 ₂₈	12 49·6 ₄₇
	12	20 58·2	20 57·4 ₂₈	12 44·9 ₄₇
	18	21 57·3	21 00·2 ₂₈	12 40·2 ₄₈
27	0	4 22 56·5	S21 03·0 ₂₈	12 12 35·4 ₄₈
Wed.	6	23 55·6	21 05·8 ₂₇	12 30·6 ₄₉
	12	24 54·8	21 08·5 ₂₈	12 25·7 ₅₀
	18	25 53·9	21 11·3 ₂₇	12 20·7 ₅₀
28	0	4 26 53·0	S21 14·0 ₂₆	12 12 15·7 ₅₀
Thur.	6	27 52·2	21 16·6 ₂₇	12 10·7 ₅₁
	12	28 51·3	21 19·3 ₂₆	12 05·6 ₅₁
	18	29 50·5	21 21·9 ₂₆	12 00·5 ₅₁
29	0	4 30 49·6	S21 24·5 ₂₅	12 11 55·4 ₅₃
Fri.	6	31 48·7	21 27·0 ₂₆	11 50·1 ₅₂
	12	32 47·9	21 29·6 ₂₅	11 44·9 ₅₃
	18	33 47·0	21 32·1 ₂₅	11 39·6 ₅₄
30	0	4 34 46·1	S21 34·6 ₂₅	12 11 34·2 ₅₃
Sat.	6	35 45·3	21 37·1 ₂₄	11 28·9 ₅₅
	12	36 44·4	21 39·5 ₂₄	11 23·4 ₅₄
	18	37 43·6	21 41·9 ₂₄	11 18·0 ₅₆
31	0	4 38 42·7	S21 44·3 ₂₄	12 11 12·4 ₅₅
Sun.	6	39 41·8	21 46·7 ₂₃	11 06·9 ₅₆
	12	40 41·0	21 49·0 ₂₃	11 01·3 ₅₆
	18	41 40·1	21 51·3 ₂₃	10 55·7 ₅₇
	24	4 42 39·2	S21 53·6	12 10 50·0

Sun's SD　16′·2

SUNSET

Date	South Latitude									North Latitude							
	60°	55°	50°	45°	40°	30°	20°	10°	0°	10°	20°	30°	40°	45°	50°	55°	60°
Nov.	h	h	h	h	h	h	h	h	h	h	h	h	h	h	h	h	h
1	19·6	19·3	19·0	18·8	18·6	18·4	18·1	18·0	17·8	17·6	17·4	17·2	17·0	16·8	16·6	16·4	16·1
6	19·9	19·4	19·2	18·9	18·7	18·4	18·2	18·0	17·8	17·6	17·4	17·2	16·9	16·7	16·5	16·2	15·9
11	20·1	19·6	19·3	19·1	18·8	18·5	18·2	18·0	17·8	17·6	17·4	17·1	16·8	16·6	16·4	16·1	15·7
16	20·3	19·8	19·4	19·2	18·9	18·6	18·3	18·0	17·8	17·6	17·3	17·1	16·7	16·5	16·2	15·9	15·5
21	20·5	20·0	19·6	19·3	19·0	18·6	18·3	18·1	17·8	17·6	17·3	17·0	16·7	16·4	16·2	15·8	15·3
26	20·7	20·1	19·7	19·4	19·1	18·7	18·4	18·1	17·9	17·6	17·3	17·0	16·6	16·4	16·1	15·7	15·2
31	20·9	20·3	19·8	19·5	19·2	18·8	18·4	18·1	17·9	17·6	17·3	17·0	16·6	16·3	16·0	15·6	15·1

Moon's Phases:　Full Moon 20ᵈ 01ʰ 34ᵐ　　Last Quarter 27ᵈ 15ʰ 46ᵐ

UT d h	R h m s	Dec ° ′	E h m s	UT d h	R h m s	Dec ° ′	E h m s
1 0	4 38 42·7	S21 44·3 ₂₄	12 11 12·4 ₅₅	**9** 0	5 10 15·2	S22 46·7 ₁₅	12 07 56·2 ₆₇
Sun. 6	39 41·8	21 46·7 ₂₃	11 06·9 ₅₆	**Mon.** 6	11 14·3	22 48·2 ₁₄	07 49·5 ₆₇
12	40 41·0	21 49·0 ₂₃	11 01·3 ₅₆	12	12 13·5	22 49·6 ₁₅	07 42·8 ₆₇
18	41 40·1	21 51·3 ₂₃	10 55·7 ₅₇	18	13 12·6	22 51·1 ₁₄	07 36·1 ₆₈
2 0	4 42 39·2	S21 53·6 ₂₃	12 10 50·0 ₅₇	**10** 0	5 14 11·7	S22 52·5 ₁₄	12 07 29·3 ₆₇
Mon. 6	43 38·4	21 55·9 ₂₂	10 44·3 ₅₈	**Tues.** 6	15 10·9	22 53·9 ₁₃	07 22·6 ₆₈
12	44 37·5	21 58·1 ₂₂	10 38·5 ₅₈	12	16 10·0	22 55·2 ₁₃	07 15·8 ₆₉
18	45 36·7	22 00·3 ₂₂	10 32·7 ₅₈	18	17 09·2	22 56·5 ₁₃	07 08·9 ₆₈
3 0	4 46 35·8	S22 02·5 ₂₁	12 10 26·9 ₅₉	**11** 0	5 18 08·3	S22 57·8 ₁₃	12 07 02·1 ₆₉
Tues. 6	47 34·9	22 04·6 ₂₂	10 21·0 ₅₉	**Wed.** 6	19 07·4	22 59·1 ₁₃	06 55·2 ₆₉
12	48 34·1	22 06·8 ₂₁	10 15·1 ₆₀	12	20 06·6	23 00·4 ₁₂	06 48·3 ₆₉
18	49 33·2	22 08·9 ₂₀	10 09·1 ₆₀	18	21 05·7	23 01·6 ₁₂	06 41·4 ₆₉
4 0	4 50 32·4	S22 10·9 ₂₁	12 10 03·1 ₆₀	**12** 0	5 22 04·9	S23 02·8 ₁₁	12 06 34·5 ₇₀
Wed. 6	51 31·5	22 13·0 ₂₀	09 57·1 ₆₁	**Thur.** 6	23 04·0	23 03·9 ₁₁	06 27·5 ₇₀
12	52 30·7	22 15·0 ₂₀	09 51·0 ₆₁	12	24 03·1	23 05·0 ₁₁	06 20·5 ₇₀
18	53 29·8	22 17·0 ₁₉	09 44·9 ₆₁	18	25 02·3	23 06·1 ₁₁	06 13·5 ₇₀
5 0	4 54 28·9	S22 18·9 ₂₀	12 09 38·8 ₆₂	**13** 0	5 26 01·4	S23 07·2 ₁₀	12 06 06·5 ₇₀
Thur. 6	55 28·1	22 20·9 ₁₉	09 32·6 ₆₂	**Fri.** 6	27 00·5	23 08·2 ₁₁	05 59·5 ₇₁
12	56 27·2	22 22·8 ₁₉	09 26·4 ₆₂	12	27 59·7	23 09·3 ₉	05 52·4 ₇₁
18	57 26·4	22 24·7 ₁₈	09 20·2 ₆₃	18	28 58·8	23 10·2 ₁₀	05 45·3 ₇₁
6 0	4 58 25·5	S22 26·5 ₁₉	12 09 13·9 ₆₃	**14** 0	5 29 58·0	S23 11·2 ₉	12 05 38·2 ₇₁
Fri. 6	4 59 24·6	22 28·4 ₁₈	09 07·6 ₆₄	**Sat.** 6	30 57·1	23 12·1 ₉	05 31·1 ₇₁
12	5 00 23·8	22 30·2 ₁₇	09 01·2 ₆₃	12	31 56·2	23 13·0 ₉	05 24·0 ₇₂
18	01 22·9	22 31·9 ₁₈	08 54·9 ₆₄	18	32 55·4	23 13·9 ₈	05 16·8 ₇₁
7 0	5 02 22·1	S22 33·7 ₁₇	12 08 48·5 ₆₅	**15** 0	5 33 54·5	S23 14·7 ₈	12 05 09·7 ₇₂
Sat. 6	03 21·2	22 35·4 ₁₇	08 42·0 ₆₄	**Sun.** 6	34 53·7	23 15·5 ₈	05 02·5 ₇₂
12	04 20·4	22 37·1 ₁₇	08 35·6 ₆₅	12	35 52·8	23 16·3 ₈	04 55·3 ₇₂
18	05 19·5	22 38·8 ₁₆	08 29·1 ₆₆	18	36 51·9	23 17·1 ₇	04 48·1 ₇₃
8 0	5 06 18·6	S22 40·4 ₁₆	12 08 22·5 ₆₅	**16** 0	5 37 51·1	S23 17·8 ₇	12 04 40·8 ₇₂
Sun. 6	07 17·8	22 42·0 ₁₆	08 16·0 ₆₆	**Mon.** 6	38 50·2	23 18·5 ₆	04 33·6 ₇₃
12	08 16·9	22 43·6 ₁₅	08 09·4 ₆₆	12	39 49·3	23 19·1 ₇	04 26·3 ₇₂
18	09 16·1	22 45·1 ₁₆	08 02·8 ₆₆	18	40 48·5	23 19·8 ₆	04 19·1 ₇₃
24	5 10 15·2	S22 46·7	12 07 56·2	24	5 41 47·6	S23 20·4	12 04 11·8

Sun's SD   16′·3

SUNRISE

Date	South Latitude 60°	55°	50°	45°	40°	30°	20°	10°	0°	North Latitude 10°	20°	30°	40°	45°	50°	55°	60°
Dec.	h	h	h	h	h	h	h	h	h	h	h	h	h	h	h	h	h
1	2·7	3·4	3·8	4·2	4·4	4·8	5·2	5·5	5·7	6·0	6·3	6·6	7·0	7·3	7·6	8·0	8·6
6	2·6	3·3	3·8	4·1	4·4	4·8	5·2	5·5	5·8	6·1	6·4	6·7	7·1	7·4	7·7	8·1	8·7
11	2·5	3·3	3·7	4·1	4·4	4·9	5·2	5·5	5·8	6·1	6·4	6·7	7·2	7·5	7·8	8·2	8·9
16	2·5	3·2	3·7	4·1	4·4	4·9	5·2	5·6	5·9	6·2	6·5	6·8	7·2	7·5	7·9	8·3	9·0
21	2·5	3·3	3·8	4·2	4·5	4·9	5·3	5·6	5·9	6·2	6·5	6·9	7·3	7·6	7·9	8·4	9·0
26	2·6	3·3	3·8	4·2	4·5	5·0	5·3	5·7	6·0	6·2	6·5	6·9	7·3	7·6	8·0	8·4	9·1
31	2·7	3·4	3·9	4·3	4·6	5·0	5·4	5·7	6·0	6·3	6·6	6·9	7·4	7·6	8·0	8·4	9·1

Moon's Phases: New Moon 4ᵈ 07ʰ 34ᵐ   First Quarter 11ᵈ 15ʰ 49ᵐ

UT		R	Dec	E	UT		R	Dec	E
d	h	h m s	° ′	h m s	d	h	h m s	° ′	h m s
17	0	5 41 47·6	S23 20·4 $_6$	12 04 11·8 $_{73}$	25	0	6 13 20·1	S23 24·3 $_3$	12 00 14·8 $_{74}$
Tues.	6	42 46·8	23 21·0 $_5$	04 04·5 $_{73}$	Wed.	6	14 19·3	23 24·0 $_4$	00 07·4 $_{74}$
	12	43 45·9	23 21·5 $_5$	03 57·2 $_{73}$		12	15 18·4	23 23·6 $_4$	12 00 00·0 $_{75}$
	18	44 45·0	23 22·0 $_5$	03 49·9 $_{74}$		18	16 17·5	23 23·2 $_5$	11 59 52·5 $_{74}$
18	0	5 45 44·2	S23 22·5 $_5$	12 03 42·5 $_{73}$	26	0	6 17 16·7	S23 22·7 $_5$	11 59 45·1 $_{74}$
Wed.	6	46 43·3	23 23·0 $_4$	03 35·2 $_{73}$	Thur.	6	18 15·8	23 22·2 $_5$	59 37·7 $_{74}$
	12	47 42·5	23 23·4 $_4$	03 27·9 $_{74}$		12	19 15·0	23 21·7 $_5$	59 30·3 $_{74}$
	18	48 41·6	23 23·8 $_4$	03 20·5 $_{74}$		18	20 14·1	23 21·2 $_6$	59 22·9 $_{74}$
19	0	5 49 40·7	S23 24·2 $_3$	12 03 13·1 $_{73}$	27	0	6 21 13·2	S23 20·6 $_6$	11 59 15·5 $_{74}$
Thur.	6	50 39·9	23 24·5 $_3$	03 05·8 $_{73}$	Fri.	6	22 12·4	23 20·0 $_6$	59 08·1 $_{74}$
	12	51 39·0	23 24·8 $_3$	02 58·4 $_{74}$		12	23 11·5	23 19·4 $_7$	59 00·7 $_{74}$
	18	52 38·2	23 25·1 $_3$	02 51·0 $_{74}$		18	24 10·6	23 18·7 $_6$	58 53·3 $_{74}$
20	0	5 53 37·3	S23 25·4 $_2$	12 02 43·6 $_{74}$	28	0	6 25 09·8	S23 18·1 $_8$	11 58 45·9 $_{74}$
Fri.	6	54 36·5	23 25·6 $_2$	02 36·2 $_{74}$	Sat.	6	26 08·9	23 17·3 $_7$	58 38·5 $_{73}$
	12	55 35·6	23 25·8 $_2$	02 28·8 $_{75}$		12	27 08·1	23 16·6 $_8$	58 31·2 $_{74}$
	18	56 34·7	23 26·0 $_1$	02 21·3 $_{74}$		18	28 07·2	23 15·8 $_8$	58 23·8 $_{73}$
21	0	5 57 33·9	S23 26·1 $_1$	12 02 13·9 $_{74}$	29	0	6 29 06·3	S23 15·0 $_8$	11 58 16·5 $_{73}$
Sat.	6	58 33·0	23 26·2 $_1$	02 06·5 $_{74}$	Sun.	6	30 05·5	23 14·2 $_9$	58 09·2 $_{73}$
	12	5 59 32·2	23 26·3 $_0$	01 59·1 $_{75}$		12	31 04·6	23 13·3 $_9$	58 01·9 $_{73}$
	18	6 00 31·3	23 26·3 $_1$	01 51·6 $_{74}$		18	32 03·8	23 12·4 $_9$	57 54·6 $_{73}$
22	0	6 01 30·4	S23 26·4 $_0$	12 01 44·2 $_{74}$	30	0	6 33 02·9	S23 11·5 $_9$	11 57 47·3 $_{73}$
Sun.	6	02 29·6	23 26·4 $_1$	01 36·8 $_{75}$	Mon.	6	34 02·0	23 10·6 $_{10}$	57 40·0 $_{73}$
	12	03 28·7	23 26·3 $_0$	01 29·3 $_{74}$		12	35 01·2	23 09·6 $_{10}$	57 32·7 $_{72}$
	18	04 27·9	23 26·3 $_1$	01 21·9 $_{75}$		18	36 00·3	23 08·6 $_{10}$	57 25·5 $_{72}$
23	0	6 05 27·0	S23 26·2 $_2$	12 01 14·4 $_{74}$	31	0	6 36 59·5	S23 07·6 $_{11}$	11 57 18·3 $_{72}$
Mon.	6	06 26·1	23 26·0 $_1$	01 07·0 $_{75}$	Tues.	6	37 58·6	23 06·5 $_{11}$	57 11·1 $_{72}$
	12	07 25·3	23 25·9 $_2$	00 59·5 $_{74}$		12	38 57·7	23 05·4 $_{11}$	57 03·9 $_{72}$
	18	08 24·4	23 25·7 $_2$	00 52·1 $_{75}$		18	39 56·9	23 04·3 $_{12}$	56 56·7 $_{72}$
24	0	6 09 23·6	S23 25·5 $_3$	12 00 44·6 $_{74}$	32	0	6 40 56·0	S23 03·1 $_{11}$	11 56 49·5 $_{71}$
Tues.	6	10 22·7	23 25·2 $_2$	00 37·2 $_{75}$	Wed.	6	41 55·2	23 02·0 $_{12}$	56 42·4 $_{72}$
	12	11 21·8	23 25·0 $_3$	00 29·7 $_{74}$		12	42 54·3	23 00·8 $_{13}$	56 35·2 $_{71}$
	18	12 21·0	23 24·7 $_3$	00 22·3 $_{75}$		18	43 53·4	22 59·5 $_{12}$	56 28·1 $_{71}$
	24	6 13 20·1	S23 24·3 $_4$	12 00 14·8 $_{75}$		24	6 44 52·6	S22 58·3	11 56 21·0 $_{71}$

Sun's SD   16′·3

SUNSET

Date	South Latitude								0°	North Latitude							
	60°	55°	50°	45°	40°	30°	20°	10°	0°	10°	20°	30°	40°	45°	50°	55°	60°
Dec.	h	h	h	h	h	h	h	h	h	h	h	h	h	h	h	h	h
1	20·9	20·3	19·8	19·5	19·2	18·8	18·4	18·1	17·9	17·6	17·3	17·0	16·6	16·3	16·0	15·6	15·1
6	21·1	20·4	19·9	19·6	19·3	18·9	18·5	18·2	17·9	17·6	17·3	17·0	16·6	16·3	16·0	15·6	15·0
11	21·2	20·5	20·0	19·7	19·4	18·9	18·6	18·2	17·9	17·7	17·4	17·0	16·6	16·3	16·0	15·5	14·9
16	21·3	20·6	20·1	19·7	19·4	19·0	18·6	18·3	18·0	17·7	17·4	17·0	16·6	16·3	16·0	15·5	14·9
21	21·4	20·6	20·1	19·8	19·5	19·0	18·6	18·3	18·0	17·7	17·4	17·1	16·6	16·4	16·0	15·6	14·9
26	21·4	20·7	20·2	19·8	19·5	19·1	18·7	18·4	18·1	17·8	17·5	17·1	16·7	16·4	16·1	15·6	14·9
31	21·4	20·7	20·2	19·8	19·5	19·1	18·7	18·4	18·1	17·8	17·5	17·2	16·7	16·5	16·1	15·7	15·1

Moon's Phases: Full Moon 19^d 19^h 10^m Last Quarter 27^d 00^h 31^m

No.	Mag.	RA	Jan.	Feb.	Mar.	Apr.	May	June	July	Aug.	Sept.	Oct.	Nov.	Dec.	Jan.
		h m	s	s	s	s	s	s	s	s	s	s	s	s	s
1	2·1	0 08	28·3	27·9	27·6	27·7	28·2	29·1	30·1	31·1	31·8	32·1	32·1	31·8	31·3
2	2·4	0 09	15·8	14·8	14·2	14·2	15·0	16·3	17·8	19·2	20·2	20·6	20·4	19·7	18·8
3	3·9	0 09	29·5	29·0	28·7	28·8	29·3	30·3	31·6	32·7	33·6	34·0	33·9	33·4	32·8
4	2·9	0 13	19·2	18·8	18·6	18·7	19·2	20·0	21·0	21·9	22·5	22·9	22·9	22·6	22·3
5	3·7	0 19	30·7	30·4	30·2	30·2	30·7	31·4	32·4	33·3	34·0	34·4	34·4	34·2	33·8
6	4·3	0 20	09·3	08·2	07·6	07·5	08·2	09·7	11·5	13·3	14·6	15·2	15·0	14·1	12·9
7	2·9	0 25	50·1	47·8	46·3	46·0	47·1	49·5	52·6	55·9	58·3	59·4	58·8	56·8	54·2
8	3·9	0 26	17·0	16·5	16·1	16·1	16·6	17·5	18·7	19·8	20·7	21·2	21·2	20·8	20·2
9	2·4	0 26	21·9	21·4	21·1	21·1	21·5	22·4	23·6	24·7	25·6	26·1	26·0	25·6	25·1
10	4·2	0 33	05·7	04·5	03·7	03·5	04·2	05·6	07·2	08·8	10·0	10·6	10·6	09·9	08·9
11	3·7	0 37	03·9	03·1	02·5	02·4	02·9	04·0	05·4	06·7	07·7	08·3	08·3	07·8	07·1
12	3·5	0 39	25·1	24·6	24·3	24·3	24·7	25·5	26·6	27·6	28·4	28·8	28·9	28·7	28·3
13	2·3	0 40	36·2	35·3	34·6	34·5	35·1	36·2	37·7	39·0	40·1	40·7	40·7	40·2	39·5
14	2·2	0 43	40·4	40·1	39·8	39·8	40·2	40·9	41·9	42·8	43·6	44·0	44·1	43·9	43·6
15	4·3	0 47	25·7	25·3	25·0	24·9	25·3	26·1	27·1	28·1	28·9	29·3	29·4	29·2	28·9
16	3·6	0 49	12·1	11·2	10·5	10·3	10·8	12·0	13·5	14·9	16·1	16·7	16·8	16·4	15·6
17	4·4	0 49	54·5	53·9	53·5	53·4	53·8	54·7	55·8	56·9	57·8	58·4	58·5	58·2	57·8
18	1−4†	0 56	48·9	47·8	47·0	46·7	47·2	48·4	50·0	51·5	52·8	53·5	53·6	53·2	52·3
19	3·9	0 56	51·0	50·4	50·0	49·9	50·3	51·1	52·2	53·3	54·2	54·7	54·9	54·7	54·2
20	4·4	0 58	41·3	40·9	40·6	40·5	40·8	41·5	42·5	43·5	44·4	44·9	45·0	44·8	44·4
21	4·4	1 03	02·0	01·6	01·3	01·3	01·6	02·3	03·2	04·1	04·9	05·4	05·5	05·4	05·1
22	3·3	1 06	09·7	09·1	08·6	08·4	08·7	09·5	10·6	11·8	12·9	13·5	13·6	13·3	12·7
23	3·6	1 08	40·6	40·3	40·0	39·9	40·2	40·8	41·8	42·7	43·5	44·0	44·2	44·0	43·8
24	2·4	1 09	49·8	49·3	48·9	48·7	49·1	49·9	50·9	52·0	52·9	53·5	53·7	53·5	53·1
25	3·8	1 24	06·7	06·3	06·0	05·9	06·1	06·7	07·6	08·6	09·4	09·9	10·1	10·1	09·8
26	2·8	1 25	56·2	55·1	54·2	53·8	54·2	55·3	56·8	58·4	59·8	60·7	61·0	60·7	59·9
27	3·4	1 28	26·7	26·1	25·6	25·3	25·5	26·2	27·2	28·4	29·4	30·1	30·3	30·1	29·6
28	3·7	1 31	34·7	34·3	34·0	33·8	34·1	34·7	35·6	36·6	37·4	38·0	38·2	38·2	38·0
29	4·0	1 31	19·8	19·1	18·4	18·1	18·3	19·0	20·1	21·4	22·5	23·2	23·5	23·2	22·6
30	4·2	1 36	54·2	53·6	53·1	52·8	53·1	53·9	54·9	56·1	57·1	57·8	58·1	58·1	57·7
31	3·8	1 38	06·3	05·6	04·9	04·6	04·9	05·7	06·9	08·2	09·3	10·1	10·5	10·4	09·9
32 d	0·6	1 37	47·3	46·3	45·5	45·0	45·1	45·9	47·2	48·6	49·9	50·8	51·0	50·6	49·8
33	4·2	1 43	46·6	45·9	45·1	44·8	45·0	45·9	47·1	48·4	49·6	50·4	50·8	50·7	50·3
34	3·6	1 44	09·1	08·7	08·3	08·1	08·3	08·9	09·7	10·7	11·5	12·1	12·3	12·3	12·0
35	3·9	1 51	33·0	32·7	32·3	32·1	32·2	32·8	33·6	34·6	35·4	36·0	36·3	36·3	36·1
36	3·6	1 53	11·2	10·7	10·2	10·0	10·2	10·8	11·8	12·8	13·8	14·4	14·8	14·8	14·6
37	3·4	1 54	32·1	30·9	29·7	29·1	29·3	30·3	31·9	33·7	35·4	36·5	37·1	36·9	36·2
38	4·4	1 53	43·4	42·7	42·1	41·7	41·8	42·4	43·4	44·6	45·7	46·5	46·8	46·6	46·1
39	2·7	1 54	44·4	44·0	43·6	43·4	43·6	44·2	45·1	46·1	47·0	47·6	48·0	48·0	47·8
40	3·7	1 56	02·2	01·4	00·6	00·2	00·2	00·8	01·9	03·2	04·4	05·3	05·6	05·4	04·8
41	3·0	1 58	50·4	49·3	48·2	47·5	47·5	48·2	49·4	51·0	52·5	53·6	54·0	53·6	52·7
42	4·2	2 00	05·6	05·1	04·7	04·5	04·6	05·1	06·0	06·9	07·8	08·5	08·8	08·8	08·5
43	3·9	2 02	08·5	08·2	07·8	07·5	07·7	08·2	09·1	10·0	10·8	11·5	11·8	11·9	11·7
44	4·1	2 03	36·5	34·6	32·8	31·8	31·9	33·2	35·4	37·9	40·2	42·0	42·8	42·5	41·4
45	2·3	2 03	60·9	60·3	59·7	59·3	59·5	60·2	61·2	62·4	63·5	64·3	64·7	64·8	64·5
46	2·2	2 07	16·7	16·2	15·8	15·5	15·7	16·3	17·2	18·2	19·1	19·8	20·2	20·2	20·1
47	3·1	2 09	39·3	38·8	38·3	37·9	38·1	38·7	39·7	40·8	41·8	42·6	43·0	43·1	42·8
48	3·8	2 16	35·1	34·3	33·5	33·0	32·9	33·4	34·4	35·7	36·9	37·8	38·2	38·1	37·5
49	4·1	2 17	25·6	25·1	24·6	24·2	24·3	24·9	25·9	27·0	28·0	28·8	29·2	29·3	29·1
50	2−10*	2 19	26·4	26·0	25·6	25·4	25·4	25·9	26·7	27·6	28·5	29·2	29·6	29·7	29·5

The figures given refer to the beginning of the month, and should be interpolated to the actual date by means of the table on page 73.

No.		Name	Dec	Jan.	F.	M.	Apr.	M.	J.	July	A.	S.	Oct.	N.	D.	Jan.
			° ′	″	″	″	″	″	″		″	″	″	″	″	″
1	α	Andromedae	N 29 05	67	64	59	55	53	55	60	67	75	82	87	89	88
2	β	Cassiopeiae	N 59 09	49	46	38	31	25	23	26	33	43	53	62	68	70
3	ε	Phoenicis	S 45 43	94	91	85	77	67	58	52	51	53	59	66	71	73
4	γ	Pegasi	N 15 11	38	35	32	31	31	35	40	47	53	58	60	60	59
5	ι	Ceti	S 8 48	57	59	58	56	51	45	38	33	30	30	32	34	37
6	ζ	Tucanae	S 64 51	73	69	61	50	39	30	24	23	28	35	44	50	51
7	β	Hydri	S 77 14	62	57	48	37	26	16	11	11	17	25	34	40	41
8	κ	Phoenicis	S 43 39	90	88	83	75	65	56	50	47	49	54	61	67	69
9	α	Phoenicis	S 42 17	64	63	58	50	40	31	25	22	24	29	36	42	44
10	κ	Cassiopeiae	N 62 56	45	43	36	28	21	19	20	27	36	47	56	63	66
11	ζ	Cassiopeiae	N 53 54	38	35	29	22	16	14	17	23	32	42	51	57	59
12	δ	Andromedae	N 30 52	22	19	14	10	08	09	13	20	27	34	40	42	42
13	α	Cassiopeiae	N 56 32	64	61	55	48	42	39	41	48	57	67	76	82	85
14	β	Ceti	S 17 58	47	47	46	42	36	28	21	16	15	16	19	23	26
15	ζ	Andromedae	N 24 16	42	39	35	32	31	33	37	44	51	57	61	63	62
16	η	Cassiopeiae	N 57 49	46	43	37	30	23	21	22	28	37	47	56	63	66
17	ν	Andromedae	N 41 05	29	27	22	16	12	11	15	21	29	37	44	49	50
18†	γ	Cassiopeiae	N 60 43	50	48	42	35	28	25	26	31	40	50	60	67	71
19	μ	Andromedae	N 38 30	42	40	35	30	26	26	29	35	43	51	57	61	63
20	α	Sculptoris	S 29 20	66	66	63	57	49	41	33	29	28	31	36	42	45
21	ε	Piscium	N 7 53	58	56	54	54	55	60	65	71	76	80	81	80	79
22	β	Phoenicis	S 46 42	51	50	45	37	27	17	10	06	07	13	20	27	31
23	η	Ceti	S 10 10	29	31	30	28	23	16	10	04	01	01	03	06	09
24	β	Andromedae	N 35 37	56	54	50	45	42	42	45	51	58	66	72	76	77
25	θ	Ceti	S 8 10	34	35	35	33	29	22	16	10	07	06	08	11	14
26	δ	Cassiopeiae	N 60 14	55	54	49	42	35	31	32	36	44	54	63	71	75
27	γ	Phoenicis	S 43 18	50	50	46	39	29	19	11	06	07	11	18	26	30
28	η	Piscium	N 15 21	20	18	16	14	14	17	22	27	33	37	40	40	40
29	δ	Phoenicis	S 49 03	67	66	62	54	44	33	25	21	22	27	34	42	47
30	υ	Andromedae	N 41 24	62	61	57	51	47	45	47	52	59	67	74	79	81
31	51	Andromedae	N 48 38	26	25	21	15	09	07	08	12	20	28	36	43	46
32		Achernar (α Eri)	S 57 13	59	59	54	45	34	23	15	11	12	18	27	35	39
33	φ	Persei	N 50 41	64	64	60	53	48	45	45	50	57	65	74	80	84
34	τ	Ceti	S 15 55	50	51	50	47	41	34	26	21	18	18	22	26	29
35	ζ	Ceti	S 10 19	42	43	43	41	36	29	22	17	13	13	15	19	22
36	α	Trianguli	N 29 35	21	20	17	14	11	11	14	19	25	31	36	39	41
37	ε	Cassiopeiae	N 63 40	58	59	55	48	41	36	35	38	45	54	64	73	78
38	ψ	Phoenicis	S 46 17	55	56	52	45	35	25	16	11	11	15	23	31	36
39	β	Arietis	N 20 48	64	62	60	58	57	58	62	67	73	78	81	83	83
40	χ	Eridani	S 51 35	78	79	75	67	57	46	37	32	32	37	45	53	59
41	α	Hydri	S 61 33	60	60	55	47	36	25	16	11	12	18	27	35	41
42	υ	Ceti	S 21 03	80	81	80	76	70	62	54	48	46	47	51	56	60
43	α	Piscium	N 2 46	18	16	15	15	18	22	28	34	38	40	40	39	37
44	50	Cassiopeiae	N 72 25	63	65	62	54	46	40	38	40	46	56	66	76	82
45	γ	Andromedae	N 42 20	27	27	24	19	15	12	14	18	24	31	38	44	47
46	α	Arietis	N 23 28	19	18	16	13	12	13	16	21	27	32	35	38	38
47	β	Trianguli	N 34 59	52	52	49	45	41	40	42	47	53	59	65	69	71
48	φ	Eridani	S 51 29	92	93	90	83	72	62	52	47	46	51	59	67	73
49	γ	Trianguli	N 33 51	27	26	24	20	17	16	18	22	28	34	39	43	45
50*	ο	Ceti	S 2 57	75	76	77	76	73	67	61	55	51	50	51	53	56

† **18:** mag. 2·2 (2000)

* **50:** At maximum (mag. 2–5) in August.

No.	Mag.	RA	Jan.	Feb.	Mar.	Apr.	May	June	July	Aug.	Sept.	Oct.	Nov.	Dec.	Jan.
		h m	s	s	s	s	s	s	s	s	s	s	s	s	s
51	4·3	2 21	48·5	46·8	45·3	44·2	43·8	44·4	45·8	47·8	49·7	51·1	51·7	51·3	50·0
52	4·4	2 27	03·8	03·0	02·3	01·8	01·7	02·1	03·1	04·2	05·4	06·3	06·7	06·7	06·2
53	4·3	2 28	15·5	15·2	14·7	14·5	14·5	15·0	15·8	16·7	17·6	18·4	18·8	18·9	18·8
54	4·0	2 39	34·8	34·5	34·0	33·7	33·7	34·2	34·9	35·8	36·7	37·5	37·9	38·1	38·0
55	4·3	2 39	38·9	37·3	35·7	34·5	34·0	34·4	35·7	37·5	39·5	41·0	41·7	41·4	40·2
56	4·1	2 40	44·9	44·3	43·7	43·2	43·1	43·5	44·3	45·4	46·5	47·3	47·8	47·8	47·5
57	3·6	2 43	24·0	23·6	23·2	22·9	22·9	23·3	24·1	25·0	25·9	26·6	27·1	27·3	27·2
58	4·2	2 44	20·0	19·4	18·6	18·1	18·0	18·7	19·7	21·0	22·3	23·4	24·1	24·3	24·1
59	4·4	2 44	12·9	12·5	12·0	11·7	11·7	12·1	12·8	13·7	14·7	15·4	15·8	16·0	15·8
60	4·4	2 45	02·7	02·4	01·9	01·6	01·6	02·1	02·8	03·8	04·7	05·4	05·9	06·1	06·0
61	3·7	2 50	05·8	05·4	04·9	04·5	04·5	05·0	05·8	06·8	07·9	08·7	09·2	09·4	09·4
62	3·9	2 50	50·6	49·8	48·8	48·2	48·1	48·7	49·9	51·4	52·9	54·1	54·9	55·1	54·9
63	4·1	2 54	23·9	23·2	22·4	21·7	21·7	22·3	23·4	24·7	26·1	27·3	28·1	28·3	28·1
64	4·0	2 56	31·4	31·0	30·5	30·2	30·1	30·5	31·2	32·1	33·0	33·8	34·3	34·5	34·4
65	3·4	2 58	20·5	20·0	19·3	18·7	18·6	18·9	19·7	20·7	21·8	22·7	23·2	23·3	23·0
66	2·8	3 02	22·8	22·5	22·1	21·7	21·7	22·1	22·8	23·7	24·6	25·4	25·9	26·1	26·1
67	4·2	3 02	28·8	28·4	27·9	27·4	27·3	27·7	28·4	29·3	30·3	31·1	31·6	31·7	31·6
68	3·1	3 04	56·5	55·8	54·9	54·3	54·1	54·7	55·8	57·2	58·6	59·8	60·7	61·0	60·8
69	3·7	3 05	18·1	17·7	17·0	16·6	16·5	17·0	17·9	19·0	20·1	21·1	21·7	22·0	22·0
70	2−3	3 08	17·8	17·3	16·7	16·2	16·1	16·6	17·5	18·6	19·8	20·8	21·5	21·8	21·7
71	4·2	3 09	12·7	12·1	11·3	10·7	10·6	11·1	12·1	13·4	14·8	15·9	16·7	17·1	17·0
72	3·9	3 12	09·5	09·1	08·5	08·0	07·9	08·2	08·9	09·8	10·8	11·6	12·2	12·4	12·2
73	3·9	3 19	36·5	36·1	35·6	35·1	34·9	35·2	35·9	36·8	37·7	38·6	39·1	39·3	39·3
74	4·3	3 19	61·0	60·4	59·7	59·1	58·8	59·1	59·8	60·9	62·0	63·0	63·6	63·8	63·6
75	1·9	3 24	28·0	27·4	26·6	26·0	25·8	26·3	27·2	28·5	29·9	31·0	31·9	32·3	32·3
76	3·8	3 24	55·1	54·8	54·4	54·0	53·9	54·2	54·9	55·8	56·7	57·5	58·1	58·4	58·4
77	3·7	3 27	16·6	16·3	15·8	15·4	15·3	15·6	16·3	17·2	18·1	18·9	19·5	19·8	19·9
78	4·4	3 29	14·2	13·4	12·3	11·4	11·1	11·6	12·8	14·3	16·0	17·5	18·6	19·1	19·0
79	4·3	3 30	58·9	58·6	58·1	57·7	57·6	57·9	58·6	59·5	60·5	61·3	61·9	62·2	62·3
80	3·8	3 33	01·5	01·2	00·7	00·3	00·1	00·4	01·0	01·9	02·8	03·6	04·2	04·5	04·4
81	4·3	3 33	52·8	52·4	51·8	51·4	51·2	51·4	52·0	52·9	53·9	54·7	55·3	55·6	55·5
82	4·4	3 36	58·5	58·2	57·7	57·3	57·1	57·4	58·1	58·9	59·8	60·7	61·3	61·6	61·6
83	3·1	3 43	04·2	03·7	02·9	02·3	02·0	02·4	03·3	04·5	05·8	07·0	07·9	08·4	08·4
84	3·7	3 43	20·7	20·4	19·9	19·5	19·3	19·6	20·2	21·0	21·9	22·7	23·4	23·7	23·7
85	3·9	3 44	26·6	26·3	25·7	25·2	25·1	25·4	26·1	27·1	28·2	29·2	29·9	30·4	30·4
86	3·8	3 44	59·6	59·3	58·8	58·3	58·2	58·5	59·2	60·1	61·1	62·0	62·7	63·1	63·2
87	3·9	3 45	19·8	19·4	18·8	18·2	17·9	18·3	19·1	20·2	21·4	22·6	23·4	23·9	23·9
88	3·8	3 44	15·7	14·4	12·9	11·6	10·7	10·7	11·5	12·9	14·6	16·1	17·1	17·3	16·7
89	4·3	3 46	56·3	55·9	55·4	54·9	54·6	54·8	55·4	56·3	57·2	58·1	58·7	59·0	59·0
90	3·0	3 47	36·2	35·9	35·4	34·9	34·7	35·0	35·7	36·6	37·6	38·6	39·3	39·7	39·8
91	3·2	3 47	16·7	14·5	12·0	09·8	08·2	07·9	08·9	10·9	13·4	15·7	17·2	17·4	16·2
92	3·8	3 49	16·8	16·6	16·0	15·6	15·4	15·7	16·4	17·3	18·3	19·2	19·9	20·4	20·5
93	4·2	3 49	32·3	31·8	31·1	30·5	30·2	30·3	30·9	31·8	32·9	33·8	34·5	34·8	34·7
94	2·9	3 54	15·5	15·2	14·6	14·1	13·9	14·2	14·9	15·9	16·9	17·9	18·7	19·2	19·3
95	3·0	3 57	59·4	59·0	58·4	57·8	57·5	57·8	58·6	59·6	60·8	61·9	62·8	63·3	63·4
96	3·2	3 58	07·5	07·2	06·7	06·3	06·0	06·2	06·8	07·6	08·5	09·4	10·0	10·4	10·4
97	4·0	3 59	05·7	05·4	04·8	04·3	04·0	04·3	05·1	06·1	07·2	08·2	09·1	09·5	09·7
98	4·4	3 58	48·6	47·5	46·2	45·0	44·2	44·1	44·7	46·0	47·5	48·9	49·9	50·2	49·8
99	3·9	4 00	47·5	47·2	46·8	46·3	46·1	46·3	47·0	47·8	48·8	49·6	50·3	50·8	50·9
100	3·9	4 03	15·8	15·6	15·1	14·6	14·4	14·7	15·2	16·1	17·0	17·9	18·5	19·0	19·1

The figures given refer to the beginning of the month, and should be interpolated to the actual date by means of the table on page 73.

No.		Name	Dec	Jan.	F.	M.	Apr.	M.	J.	July	A.	S.	Oct.	N.	D.	Jan.
			° ′	″	″	″	″	″	″	″	″	″	″	″	″	″
51	δ	Hydri	S 68 38	84	85	81	72	61	50	40	35	36	42	51	60	66
52	κ	Eridani	S 47 41	62	63	61	54	44	34	24	18	17	21	29	37	43
53	ξ²	Ceti	N 8 28	04	03	01	01	02	06	11	16	21	23	25	24	22
54	δ	Ceti	N 0 20	07	05	04	05	08	12	18	24	28	30	29	27	25
55	ε	Hydri	S 68 15	52	54	50	42	31	20	10	05	04	10	18	28	35
56	ι	Eridani	S 39 50	66	69	67	62	53	43	33	27	25	28	35	42	49
57	γ	Ceti	N 3 14	33	32	31	31	33	37	43	48	52	55	55	53	51
58	θ	Persei	N 49 14	20	21	19	14	09	05	04	07	12	19	26	33	37
59	π	Ceti	S 13 50	72	74	74	72	67	60	52	46	43	42	45	50	54
60	μ	Ceti	N 10 07	18	16	15	14	15	18	23	28	32	35	37	36	35
61	41	Arietis	N 27 16	09	09	07	04	02	02	04	08	13	18	22	25	26
62	η	Persei	N 55 54	22	24	22	17	11	06	04	05	10	17	26	33	39
63	τ	Persei	N 52 46	22	24	22	17	11	07	05	07	12	19	26	34	39
64	η	Eridani	S 8 53	34	36	36	35	31	25	18	12	08	07	09	13	17
65*	θ	Eridani	S 40 17	65	68	67	61	53	43	33	27	24	27	33	41	48
66	α	Ceti	N 4 05	46	44	43	44	45	49	55	60	64	66	66	65	62
67	τ³	Eridani	S 23 36	73	75	75	72	66	57	49	42	39	40	44	51	56
68	γ	Persei	N 53 30	59	61	60	55	49	45	43	44	48	55	62	70	75
69	ρ	Persei	N 38 50	57	58	57	53	49	47	47	50	54	60	65	70	73
70		Algol (β Persei)	N 40 57	53	54	53	49	45	42	42	44	49	54	60	65	69
71	ι	Persei	N 49 37	22	24	23	18	13	09	07	09	13	19	26	33	38
72	α	Fornacis	S 28 58	58	62	61	57	51	42	33	26	23	24	29	36	42
73	16	Eridani	S 21 44	74	77	77	74	68	60	52	45	42	42	46	52	58
74	BS 1008 (Eridani)		S 43 03	60	64	63	58	49	39	29	22	19	21	28	36	44
75	α	Persei	N 49 51	72	75	74	70	65	61	59	59	63	69	76	82	87
76	o	Tauri	N 9 02	06	04	03	03	04	07	11	16	20	23	23	22	21
77	ξ	Tauri	N 9 44	20	18	17	17	18	20	25	29	33	36	37	36	35
78	BS 1035 (Cam)		N 59 56	58	62	61	57	51	45	41	41	44	50	58	66	73
79	5	Tauri	N 12 56	34	33	32	32	32	34	38	42	46	49	50	50	49
80	ε	Eridani	S 9 26	74	76	77	75	71	65	59	52	49	48	50	54	59
81	τ⁵	Eridani	S 21 37	45	49	49	46	41	33	25	18	14	15	18	25	31
82	10	Tauri	N 0 24	23	21	20	21	23	27	33	38	42	44	43	40	37
83	δ	Persei	N 47 47	44	47	46	43	38	34	32	32	35	40	47	53	58
84	δ	Eridani	S 9 45	32	35	35	34	30	24	17	11	07	06	08	12	17
85	o	Persei	N 32 17	43	44	43	41	38	37	37	39	43	47	51	54	57
86	17	Tauri	N 24 07	11	11	11	09	08	07	09	12	16	19	22	24	25
87	ν	Persei	N 42 35	10	12	12	09	05	01	00	01	04	09	14	19	24
88	β	Reticuli	S 64 47	80	84	83	78	68	57	46	39	36	39	47	57	65
89	τ⁶	Eridani	S 23 14	48	52	53	50	45	37	29	22	18	18	22	29	35
90	η	Tauri	N 24 06	41	41	41	39	38	38	39	42	46	49	52	54	55
91	γ	Hydri	S 74 13	75	80	79	73	64	52	42	34	32	35	43	53	62
92	27	Tauri	N 24 03	35	35	34	33	31	31	33	36	39	43	45	47	48
93	BS 1195 (Eridani)		S 36 11	52	56	57	53	46	37	27	20	16	17	23	31	38
94	ζ	Persei	N 31 53	24	26	25	23	21	19	19	21	25	28	32	35	38
95	ε	Persei	N 40 00	61	63	63	61	57	54	53	54	57	61	66	70	74
96	γ	Eridani	S 13 29	78	82	83	81	77	71	63	57	53	52	55	60	65
97	ξ	Persei	N 35 47	51	53	53	50	47	45	45	46	49	53	57	61	64
98	δ	Reticuli	S 61 23	56	61	61	56	48	37	26	18	14	17	24	34	43
99	λ	Tauri	N 12 29	43	42	42	41	42	43	47	51	54	57	58	57	56
100	ν	Tauri	N 5 59	38	36	35	35	37	40	44	49	53	54	54	53	50

* No., mag., dist. and p.a. of companion star: **65**, 4·4, 8″, 90°

No.	Mag.	RA	Jan.	Feb.	Mar.	Apr.	May	June	July	Aug.	Sept.	Oct.	Nov.	Dec.	Jan.
		h m	s	s	s	s	s	s	s	s	s	s	s	s	s
101	4·3	4 06	44·3	43·8	43·0	42·3	41·9	42·2	43·0	44·2	45·6	46·9	48·0	48·6	48·7
102	4·0	4 08	48·6	48·2	47·5	46·8	46·5	46·7	47·5	48·7	50·0	51·2	52·3	52·9	53·0
103	4·1	4 11	58·0	57·7	57·2	56·7	56·5	56·7	57·2	58·0	58·9	59·8	60·5	60·9	61·0
104	4·3	4 15	02·9	02·5	01·8	01·1	00·7	01·0	01·7	02·9	04·2	05·5	06·6	07·2	07·4
105	3·8	4 14	05·0	04·5	03·7	02·9	02·5	02·5	03·0	03·9	05·0	06·0	06·8	07·2	07·1
106	3·4	4 14	29·2	28·2	26·8	25·5	24·5	24·3	24·9	26·1	27·6	29·1	30·2	30·6	30·2
107	4·3	4 15	38·7	38·4	38·0	37·5	37·3	37·5	38·0	38·8	39·8	40·6	41·4	41·8	42·0
108	4·4	4 16	06·1	05·4	04·4	03·5	02·9	02·8	03·3	04·3	05·5	06·7	07·6	08·0	07·8
109	3·6	4 17	58·8	58·4	57·8	57·1	56·7	56·8	57·3	58·1	59·1	60·1	60·8	61·2	61·2
110	3·9	4 19	54·5	54·3	53·8	53·3	53·1	53·3	53·9	54·7	55·6	56·6	57·3	57·8	58·0
111	3·9	4 23	03·1	02·9	02·4	01·9	01·7	01·9	02·4	03·3	04·2	05·2	05·9	06·4	06·6
112	4·1	4 24	07·4	07·0	06·3	05·7	05·3	05·3	05·8	06·6	07·6	08·5	09·3	09·7	09·7
113	3·6	4 28	44·1	43·9	43·4	42·9	42·7	42·8	43·4	44·2	45·2	46·1	46·9	47·5	47·7
114	3·6	4 28	46·7	46·5	46·0	45·5	45·3	45·5	46·0	46·8	47·8	48·7	49·5	50·0	50·2
115	3·5	4 34	04·0	03·3	02·3	01·2	00·4	00·2	00·6	01·6	02·9	04·1	05·2	05·6	05·5
116 d	1·1	4 36	02·2	02·1	01·6	01·1	00·8	01·0	01·5	02·3	03·3	04·2	05·0	05·5	05·7
117	3·9	4 35	38·3	38·0	37·4	36·7	36·3	36·3	36·8	37·5	38·5	39·5	40·2	40·7	40·8
118	4·1	4 36	25·3	25·2	24·7	24·2	23·9	24·0	24·5	25·2	26·1	27·0	27·8	28·2	28·4
119	4·0	4 38	16·6	16·4	15·9	15·4	15·1	15·1	15·6	16·3	17·2	18·1	18·8	19·3	19·5
120	4·3	4 42	22·0	21·9	21·4	20·9	20·6	20·7	21·2	22·1	23·0	24·0	24·9	25·5	25·7
121	4·2	4 45	36·4	36·2	35·8	35·2	34·9	35·0	35·5	36·2	37·1	38·0	38·7	39·3	39·5
122	3·3	4 49	57·1	57·0	56·5	56·0	55·7	55·8	56·3	57·0	57·9	58·9	59·6	60·2	60·4
123	3·8	4 51	18·9	18·8	18·4	17·9	17·6	17·6	18·1	18·8	19·7	20·6	21·4	22·0	22·2
124	4·4	4 54	16·1	15·6	14·3	12·9	12·0	12·0	12·9	14·5	16·6	18·7	20·5	21·7	22·2
125	3·9	4 54	21·6	21·4	21·0	20·5	20·2	20·2	20·7	21·4	22·3	23·2	24·0	24·5	24·8
126	2·9	4 57	07·6	07·5	07·0	06·4	06·0	06·1	06·6	07·5	08·6	09·7	10·6	11·3	11·6
127	3·4	5 02	07·1	06·9	06·3	05·6	05·1	05·2	05·8	06·7	07·9	09·2	10·3	11·1	11·5
128	3·9	5 02	37·4	37·2	36·6	35·9	35·5	35·6	36·2	37·1	38·2	39·4	40·5	41·3	41·6
129	4·2	5 03	36·6	36·2	35·3	34·2	33·4	33·4	34·2	35·5	37·2	38·9	40·5	41·6	42·0
130	3·3	5 05	33·3	33·1	32·6	32·0	31·5	31·5	31·8	32·5	33·4	34·3	35·1	35·7	35·9
131	3·3	5 06	39·6	39·5	38·9	38·2	37·8	37·8	38·4	39·3	40·4	41·7	42·7	43·5	43·9
132	2·9	5 07	57·2	57·1	56·6	56·1	55·7	55·7	56·1	56·8	57·7	58·6	59·4	59·9	60·2
133	4·3	5 09	14·9	14·7	14·3	13·7	13·4	13·4	13·8	14·4	15·3	16·2	17·0	17·6	17·8
134	3·3	5 13	01·7	01·6	01·1	00·5	00·1	00·1	00·4	01·1	02·0	02·9	03·7	04·3	04·5
135 d	0·3	5 14	38·4	38·3	37·8	37·3	36·9	36·9	37·3	37·9	38·8	39·7	40·5	41·1	41·4
136 d	0·2	5 16	50·6	50·5	49·9	49·1	48·6	48·6	49·2	50·1	51·3	52·6	53·8	54·7	55·1
137	3·7	5 17	42·6	42·5	42·0	41·5	41·1	41·1	41·4	42·1	43·0	43·9	44·7	45·3	45·6
138	4·3	5 19	40·5	40·3	39·9	39·3	38·9	38·9	39·2	39·9	40·7	41·6	42·4	43·0	43·3
139	3·4	5 24	35·0	34·9	34·5	34·0	33·6	33·6	33·9	34·6	35·4	36·3	37·1	37·8	38·1
140	1·7	5 25	14·6	14·5	14·1	13·6	13·2	13·2	13·6	14·2	15·1	16·0	16·9	17·5	17·8
141	1·8	5 26	25·3	25·3	24·8	24·3	23·9	23·9	24·3	25·0	26·0	27·1	28·1	28·8	29·2
142	3·0	5 28	20·4	20·3	19·8	19·2	18·7	18·6	18·9	19·6	20·4	21·3	22·2	22·8	23·1
143	3·9	5 31	18·0	17·7	17·1	16·4	15·8	15·6	15·8	16·5	17·4	18·4	19·3	19·9	20·2
144	2·5	5 32	06·8	06·8	06·4	05·9	05·5	05·4	05·8	06·4	07·2	08·1	09·0	09·6	10·0
145	2·7	5 32	49·6	49·5	49·0	48·4	48·0	47·9	48·2	48·8	49·6	50·6	51·4	52·0	52·3
146	3·8	5 33	41·3	40·5	39·2	37·8	36·5	35·8	35·9	36·7	38·0	39·6	40·9	41·8	41·9
147	3·7	5 35	15·2	15·1	14·7	14·2	13·8	13·8	14·2	14·8	15·7	16·6	17·5	18·2	18·6
148	2·9	5 35	32·2	32·1	31·7	31·2	30·8	30·7	31·1	31·7	32·5	33·4	34·2	34·9	35·2
149	1·7	5 36	19·2	19·2	18·8	18·2	17·9	17·8	18·1	18·7	19·6	20·5	21·3	22·0	22·3
150	3·0	5 37	46·1	46·1	45·7	45·1	44·7	44·7	45·1	45·8	46·7	47·7	48·6	49·3	49·8

The figures given refer to the beginning of the month, and should be interpolated to the actual date by means of the table on page 73.

No.		Name	Dec	Jan.	F.	M.	Apr.	M.	J.	July	A.	S.	Oct.	N.	D.	Jan.
101	λ	Persei	N 50 21	29	33	33	31	26	21	18	18	20	24	30	36	41
102	48	Persei	N 47 42	69	72	73	70	66	62	59	59	61	65	70	76	81
103	o^1	Eridani	S 6 49	63	66	67	66	63	58	51	45	41	40	42	46	51
104	μ	Persei	N 48 24	56	60	61	58	54	50	47	46	48	52	57	63	68
105	α	Horologii	S 42 16	95	100	101	98	91	81	71	63	59	60	65	74	83
106	α	Reticuli	S 62 27	82	88	88	84	76	65	54	46	42	44	51	61	70
107	μ	Tauri	N 8 53	48	46	46	45	46	49	53	57	60	62	62	61	59
108	γ	Doradus	S 51 28	67	73	74	70	62	52	41	33	29	30	37	46	55
109	v^4	Eridani	S 33 47	47	52	54	51	45	36	27	19	14	15	20	28	36
110	γ	Tauri	N 15 37	55	55	54	54	54	55	58	61	64	66	67	67	67
111	δ	Tauri	N 17 32	49	49	48	47	47	48	50	53	56	59	60	60	60
112	43	Eridani	S 34 00	54	60	61	59	53	44	34	26	22	22	27	35	43
113	ε	Tauri	N 19 11	05	05	04	04	03	04	06	08	11	14	15	15	15
114	$θ^2$	Tauri	N 15 52	30	29	29	28	28	29	32	35	38	40	41	41	40
115	α	Doradus	S 55 02	39	46	47	44	37	27	16	07	02	03	10	19	29
116	*Aldebaran (α Tau)*		N 16 30	47	46	46	45	45	46	49	52	54	56	57	57	57
117	v^2	Eridani	S 30 33	39	44	46	44	39	30	21	13	09	09	13	21	29
118	ν	Eridani	S 3 20	59	62	63	62	60	56	50	45	41	39	41	45	49
119	53	Eridani	S 14 17	67	71	72	71	67	61	54	48	43	43	45	51	57
120	τ	Tauri	N 22 57	39	39	39	38	37	37	38	41	43	45	47	48	48
121	μ	Eridani	S 3 14	68	71	72	72	69	65	60	54	50	49	51	54	59
122	$π^3$	Orionis	N 6 57	50	49	48	48	49	52	55	60	63	64	64	62	59
123	$π^4$	Orionis	N 5 36	28	26	25	25	26	29	33	37	41	42	42	39	37
124	α	Camelopardalis	N 66 20	51	57	60	59	53	47	41	37	36	38	44	51	59
125	$π^5$	Orionis	N 2 26	35	33	32	32	33	37	41	46	49	51	50	47	44
126	ι	Aurigae	N 33 10	11	13	14	13	11	09	08	08	10	12	14	17	19
127	ε	Aurigae	N 43 49	37	41	42	41	38	35	32	31	31	33	36	41	45
128	ζ	Aurigae	N 41 04	46	49	50	50	47	44	41	40	41	43	46	50	54
129	β	Camelopardalis	N 60 26	47	53	55	54	50	44	38	35	34	36	41	47	54
130	ε	Leporis	S 22 21	72	78	80	79	75	68	60	53	48	47	51	57	65
131	η	Aurigae	N 41 14	16	19	21	20	17	14	12	11	11	13	16	20	23
132	β	Eridani	S 5 04	66	69	71	70	68	64	58	53	49	48	50	54	58
133	λ	Eridani	S 8 44	70	74	75	75	72	68	61	55	51	50	52	57	62
134	μ	Leporis	S 16 11	77	81	83	83	80	74	66	60	55	54	57	63	69
135	*Rigel (β Ori)*		S 8 11	62	66	67	67	64	60	54	48	44	43	45	49	55
136	*Capella (α Aur)*		N 45 59	62	67	69	68	65	61	58	56	56	57	60	64	69
137	τ	Orionis	S 6 50	36	40	41	41	39	34	28	23	19	18	20	24	29
138	λ	Leporis	S 13 10	33	38	40	40	37	31	25	18	14	13	15	21	27
139	η	Orionis	S 2 23	46	49	50	50	48	44	39	34	31	30	31	35	39
140	*Bellatrix (γ Ori)*		N 6 21	03	01	00	00	01	04	08	11	14	15	15	12	09
141	β	Tauri	N 28 36	33	35	36	35	34	33	33	33	34	35	36	37	39
142	β	Leporis	S 20 45	33	39	41	41	38	31	24	16	11	10	14	20	27
143	ε	Columbae	S 35 27	75	82	85	85	80	73	63	55	49	48	52	60	69
144	δ	Orionis	S 0 17	54	57	58	58	57	53	49	44	41	40	41	45	49
145	α	Leporis	S 17 48	79	85	87	87	84	78	71	64	59	58	61	67	74
146	β	Doradus	S 62 28	85	93	97	97	92	83	72	62	56	55	60	69	80
147*	λ	Orionis	N 9 56	06	04	04	04	04	06	09	12	15	16	15	13	11
148	ι	Orionis	S 5 54	34	38	40	39	37	33	28	22	18	17	19	24	29
149	ε	Orionis	S 1 11	65	68	70	70	68	64	60	55	51	50	52	56	60
150	ζ	Tauri	N 21 08	37	37	38	38	37	38	38	40	41	42	42	42	42

* No., mag., dist. and p.a. of companion star: **147**, 5·6, 4″, 43°

No.	Mag.	RA	Jan.	Feb.	Mar.	Apr.	May	June	July	Aug.	Sept.	Oct.	Nov.	Dec.	Jan.
		h m	s	s	s	s	s	s	s	s	s	s	s	s	s
151	3·8	5 38	51·1	51·1	50·7	50·1	49·8	49·7	50·0	50·6	51·4	52·3	53·2	53·9	54·2
152	2·7	5 39	44·2	44·0	43·4	42·7	42·1	41·9	42·1	42·7	43·6	44·6	45·5	46·2	46·4
153	2·0	5 40	51·9	51·9	51·5	50·9	50·6	50·5	50·8	51·4	52·2	53·1	54·0	54·6	55·0
154	3·8	5 44	33·4	33·3	32·8	32·2	31·7	31·6	31·8	32·4	33·2	34·1	35·0	35·7	36·0
155	3·7	5 47	03·3	03·2	02·8	02·2	01·7	01·6	01·9	02·5	03·3	04·2	05·0	05·7	06·1
156	2·2	5 47	51·5	51·5	51·0	50·5	50·1	50·0	50·2	50·8	51·6	52·5	53·4	54·1	54·4
157	3·9	5 47	21·6	21·2	20·4	19·4	18·5	18·0	18·1	18·7	19·8	21·0	22·1	22·9	23·1
158	4·4	5 49	53·9	53·4	52·4	51·2	50·2	49·6	49·7	50·3	51·4	52·8	54·0	54·8	55·1
159	4·2	5 51	38·1	38·1	37·6	36·9	36·4	36·3	36·7	37·5	38·6	39·7	40·9	41·8	42·3
160	3·2	5 51	02·8	02·6	02·0	01·3	00·7	00·4	00·6	01·1	02·0	03·0	04·0	04·7	05·0
161	3·9	5 51	25·0	25·0	24·5	23·9	23·4	23·3	23·5	24·0	24·9	25·8	26·7	27·3	27·7
162 d	0−1	5 55	17·1	17·1	16·7	16·2	15·8	15·8	16·0	16·6	17·5	18·4	19·3	20·0	20·4
163	3·8	5 56	30·3	30·2	29·8	29·2	28·8	28·6	28·9	29·4	30·2	31·1	32·0	32·7	33·1
164	4·4	5 57	37·4	37·3	36·7	36·0	35·4	35·1	35·2	35·8	36·6	37·6	38·6	39·3	39·6
165	3·9	5 59	42·2	42·2	41·5	40·6	39·9	39·7	40·1	41·0	42·4	43·9	45·4	46·6	47·3
166	2·1	5 59	41·0	41·0	40·5	39·8	39·2	39·0	39·4	40·2	41·3	42·6	43·9	44·9	45·5
167	2·7	5 59	51·8	51·9	51·4	50·8	50·3	50·2	50·5	51·2	52·3	53·4	54·5	55·4	56·0
168	4·0	5 59	13·7	13·5	12·8	12·0	11·3	10·9	11·0	11·5	12·4	13·5	14·5	15·3	15·6
169	4·3	6 04	14·8	14·8	14·5	13·9	13·5	13·4	13·7	14·3	15·2	16·2	17·2	18·0	18·5
170	4·4	6 07	41·5	41·5	41·2	40·6	40·2	40·1	40·4	41·0	41·8	42·8	43·7	44·5	45·0
171	3−4	6 14	60·2	60·3	59·9	59·4	58·9	58·8	59·1	59·7	60·5	61·5	62·5	63·3	63·9
172	4·4	6 15	30·6	30·7	30·4	29·8	29·3	29·2	29·5	30·1	31·0	32·0	33·1	34·0	34·6
173	4·4	6 19	48·9	49·0	48·3	47·2	46·3	46·0	46·3	47·2	48·7	50·4	52·0	53·5	54·3
174	3·1	6 20	24·2	24·2	23·7	23·0	22·4	22·2	22·2	22·7	23·5	24·4	25·4	26·2	26·6
175	4·0	6 22	12·1	12·1	11·6	10·9	10·2	09·9	10·0	10·4	11·2	12·2	13·2	13·9	14·4
176	3·2	6 23	05·2	05·3	05·0	04·4	04·0	03·8	04·1	04·6	05·5	06·5	07·5	08·3	08·9
177	2·0	6 22	47·8	47·8	47·5	46·9	46·4	46·2	46·3	46·8	47·5	48·4	49·3	50·1	50·5
178	4·5	6 23	52·8	52·9	52·5	52·0	51·6	51·5	51·7	52·2	52·9	53·8	54·7	55·5	56·0
179 d	−0·9	6 23	61·6	61·3	60·5	59·5	58·5	57·9	57·8	58·3	59·2	60·4	61·6	62·6	63·0
180	4·1	6 29	05·2	05·3	05·0	04·5	04·0	03·9	04·1	04·6	05·5	06·4	07·4	08·3	08·8
181	1·9	6 37	49·9	50·1	49·8	49·3	48·8	48·7	48·9	49·4	50·2	51·1	52·1	52·9	53·5
182	3·2	6 37	50·6	50·5	49·9	49·1	48·3	47·9	47·8	48·2	49·0	50·1	51·1	52·0	52·5
183	3·2	6 44	03·6	03·8	03·5	03·0	02·5	02·3	02·5	03·0	03·8	04·8	05·8	06·7	07·4
184	3·4	6 45	24·4	24·6	24·3	23·8	23·3	23·1	23·3	23·8	24·5	25·4	26·4	27·2	27·8
185 d	−1·6	6 45	14·6	14·7	14·3	13·8	13·3	13·0	13·1	13·5	14·2	15·1	16·0	16·8	17·3
186	3·3	6 48	15·4	15·1	14·0	12·6	11·2	10·2	09·9	10·2	11·2	12·7	14·2	15·4	15·9
187	3·8	6 49	55·8	55·9	55·4	54·8	54·1	53·7	53·7	54·1	54·8	55·7	56·7	57·6	58·1
188	2·8	6 49	60·8	60·7	60·0	59·1	58·1	57·4	57·3	57·6	58·4	59·6	60·8	61·8	62·3
189	3·6	6 52	55·6	55·8	55·5	54·9	54·4	54·2	54·3	54·9	55·7	56·8	57·9	58·9	59·7
190	4·2	6 54	17·4	17·6	17·3	16·8	16·3	16·0	16·1	16·5	17·2	18·0	18·9	19·7	20·3
191	1·6	6 58	43·1	43·1	42·7	42·1	41·5	41·2	41·1	41·5	42·2	43·1	44·0	44·9	45·4
192	3·7	7 01	48·7	48·8	48·4	47·8	47·2	46·8	46·8	47·2	47·8	48·7	49·7	50·5	51·1
193	3·1	7 03	07·1	07·3	06·9	06·3	05·8	05·4	05·4	05·8	06·4	07·3	08·3	09·1	09·7
194	3·9	7 04	13·9	14·2	13·9	13·4	12·9	12·7	12·9	13·3	14·0	15·0	16·0	16·9	17·6
195	4·1	7 03	51·4	51·6	51·3	50·8	50·2	50·0	50·0	50·4	51·0	51·9	52·8	53·6	54·2
196	2·0	7 08	29·1	29·2	28·9	28·3	27·7	27·3	27·3	27·6	28·3	29·1	30·1	31·0	31·5
197	3·9	7 08	48·1	47·6	46·2	44·2	42·2	40·5	39·7	39·8	41·0	42·9	44·9	46·6	47·4
198	4·1	7 11	58·3	58·5	58·3	57·8	57·4	57·1	57·2	57·6	58·2	59·0	60·0	60·8	61·5
199	4·5	7 12	38·5	38·5	37·9	37·1	36·2	35·6	35·4	35·6	36·4	37·4	38·5	39·6	40·2
200	3·8	7 14	54·3	54·4	54·1	53·5	52·9	52·6	52·5	52·8	53·5	54·3	55·3	56·2	56·8

The figures given refer to the beginning of the month, and should be interpolated to the actual date by means of the table on page 73.

No.		Name	Dec	Jan.	F.	M.	Apr.	M.	J.	July	A.	S.	Oct.	N.	D.	Jan.
			° ′	″	″	″	″	″	″	″	″	″	″	″	″	″
151	σ	Orionis	S 2 35	59	62	63	63	62	58	53	48	44	43	45	49	54
152	α	Columbae	S 34 04	28	35	39	39	35	27	18	09	04	03	06	14	23
153	ζ¹	Orionis	S 1 56	32	35	37	37	35	32	27	22	18	17	19	23	27
154	γ	Leporis	S 22 26	56	62	65	65	62	56	48	40	35	34	37	44	52
155	ζ	Leporis	S 14 49	20	25	27	27	25	20	13	06	02	01	03	09	15
156	κ	Orionis	S 9 39	71	75	78	78	75	71	65	59	55	54	56	61	67
157	β	Pictoris	S 51 03	62	70	75	75	70	62	51	42	35	34	38	47	57
158	γ	Pictoris	S 56 09	63	72	76	76	72	63	53	43	36	35	39	48	59
159	ν	Aurigae	N 39 08	57	61	63	63	62	59	57	55	54	54	56	58	61
160	β	Columbae	S 35 45	67	75	79	79	75	68	59	50	44	42	46	54	63
161	δ	Leporis	S 20 52	48	54	57	57	54	48	41	34	29	27	30	37	44
162	*Betelgeuse* (α Ori)		N 7 24	25	23	23	23	23	26	29	32	35	35	34	32	29
163	η	Leporis	S 14 09	65	70	73	73	71	66	59	53	48	47	50	55	62
164	γ	Columbae	S 35 16	63	70	75	75	71	64	55	46	40	39	42	50	59
165	δ	Aurigae	N 54 16	67	73	76	77	74	70	65	61	58	58	60	64	69
166	β	Aurigae	N 44 56	52	57	60	60	58	55	51	49	47	47	48	51	55
167	θ	Aurigae	N 37 12	45	48	51	51	50	48	45	44	43	43	44	46	48
168	η	Columbae	S 42 48	58	66	71	71	67	60	50	41	34	33	37	45	55
169	I	Geminorum	N 23 15	47	48	49	49	49	49	49	50	50	50	50	50	50
170	ν	Orionis	N 14 46	05	04	04	04	04	05	07	09	11	11	10	09	07
171	η	Geminorum	N 22 30	22	22	23	24	23	23	24	24	25	25	24	24	23
172	κ	Aurigae	N 29 29	50	52	53	54	53	52	51	51	50	50	50	50	51
173	2	Lyncis	N 59 00	37	44	49	50	48	43	37	32	28	27	28	32	38
174	ζ	Canis Majoris	S 30 03	53	61	65	66	63	57	49	41	35	33	36	43	52
175	δ	Columbae	S 33 25	76	84	89	90	87	81	73	64	58	56	59	66	75
176	μ	Geminorum	N 22 30	44	45	46	46	46	46	47	47	47	47	47	46	45
177	β	Canis Majoris	S 17 57	26	32	36	37	34	29	23	16	11	09	12	18	25
178	ε	Monocerotis	N 4 35	30	27	26	26	27	29	33	36	39	39	38	34	30
179	*Canopus* (α Car)		S 52 41	50	59	65	66	63	56	46	36	29	26	30	38	49
180	ν	Geminorum	N 20 12	38	38	39	40	40	40	41	42	42	42	41	40	39
181	γ	Geminorum	N 16 23	50	50	50	51	51	52	53	55	55	55	54	52	50
182	ν	Puppis	S 43 11	52	61	67	69	66	60	51	41	34	31	34	42	52
183	ε	Geminorum	N 25 07	44	45	46	47	47	47	47	47	46	45	44	43	43
184	ξ	Geminorum	N 12 53	36	34	34	34	35	36	38	40	41	41	39	37	34
185	*Sirius* (α CMa)		S 16 42	67	74	77	78	77	72	66	59	55	53	56	62	69
186	α	Pictoris	S 61 56	35	45	52	55	53	46	36	26	18	15	17	25	36
187	κ	Canis Majoris	S 32 30	38	47	52	54	52	46	38	30	23	21	23	30	40
188	τ	Puppis	S 50 36	60	70	76	79	77	70	61	51	44	40	43	51	61
189	θ	Geminorum	N 33 57	31	33	36	37	37	36	34	32	30	29	28	28	29
190	θ	Canis Majoris	S 12 02	28	34	37	38	36	33	27	21	17	15	18	23	30
191	ε	Canis Majoris	S 28 58	28	36	41	43	42	37	29	21	15	12	15	21	30
192	σ	Canis Majoris	S 27 55	74	82	88	90	88	83	76	68	62	59	61	68	77
193	o²	Canis Majoris	S 23 49	69	77	82	83	82	77	70	63	57	55	57	63	72
194	ζ	Geminorum	N 20 33	62	62	63	63	64	64	65	65	65	64	62	60	59
195	γ	Canis Majoris	S 15 37	70	76	80	81	80	76	70	63	59	57	59	65	72
196	δ	Canis Majoris	S 26 23	45	53	58	60	59	54	47	39	33	31	33	40	48
197*	γ²	Volantis	S 70 29	64	74	82	86	85	79	70	60	51	47	48	56	66
198	δ	Monocerotis	S 0 29	46	50	52	52	51	48	45	41	38	37	40	44	49
199	I	Puppis	S 46 45	42	53	59	63	62	56	47	38	30	26	28	35	46
200	ω	Canis Majoris	S 26 46	32	40	46	48	47	42	35	27	21	19	21	27	36

* No., mag., dist. and p.a. of companion star: **197**, 5·8, 14″, 300°

No.	Mag.	RA	Jan.	Feb.	Mar.	Apr.	May	June	July	Aug.	Sept.	Oct.	Nov.	Dec.	Jan.
		h m	s	s	s	s	s	s	s	s	s	s	s	s	s
201	2·7	7 17	13·8	13·9	13·5	12·8	12·1	11·6	11·5	11·8	12·4	13·3	14·4	15·3	15·9
202	3·6	7 18	12·7	13·0	12·8	12·3	11·9	11·6	11·7	12·1	12·8	13·7	14·6	15·6	16·3
203	4·0	7 16	53·4	53·1	51·9	50·1	48·3	46·9	46·1	46·2	47·3	48·9	50·8	52·4	53·1
204	3·5	7 20	14·8	15·1	14·9	14·4	13·9	13·7	13·8	14·2	14·9	15·8	16·8	17·7	18·5
205	2·4	7 24	11·2	11·4	11·0	10·5	09·8	09·4	09·3	09·6	10·2	11·1	12·1	13·0	13·6
206	3·9	7 25	51·3	51·7	51·5	51·0	50·4	50·2	50·2	50·6	51·4	52·3	53·4	54·4	55·2
207	3·1	7 27	15·8	16·1	15·9	15·5	15·0	14·8	14·8	15·1	15·8	16·6	17·6	18·5	19·2
208	4·2	7 29	14·7	15·1	14·9	14·4	13·8	13·5	13·6	14·0	14·7	15·7	16·8	17·9	18·7
209	3·3	7 29	18·9	19·0	18·5	17·8	17·0	16·4	16·2	16·4	17·0	18·0	19·1	20·1	20·7
210	1·6	7 34	43·9	44·3	44·1	43·6	43·0	42·7	42·8	43·2	43·9	44·8	46·0	47·0	47·8
211	4·2	7 36	03·0	03·4	03·2	02·7	02·2	01·9	02·0	02·3	03·0	03·9	05·0	06·0	06·8
212 d	0·5	7 39	24·6	24·9	24·7	24·3	23·8	23·5	23·6	23·9	24·5	25·3	26·2	27·1	27·8
213	4·1	7 41	21·0	21·2	21·1	20·6	20·1	19·8	19·8	20·0	20·6	21·4	22·3	23·2	23·9
214	3·9	7 41	52·4	52·2	50·8	48·7	46·4	44·4	43·1	43·0	44·0	45·9	48·2	50·2	51·3
215	3·7	7 44	34·3	34·7	34·6	34·1	33·6	33·3	33·3	33·7	34·3	35·2	36·2	37·2	38·1
216 d	1·2	7 45	26·5	26·9	26·8	26·3	25·8	25·5	25·5	25·8	26·5	27·4	28·5	29·5	30·3
217	3·7	7 45	20·5	20·7	20·4	19·8	19·0	18·5	18·3	18·5	19·0	19·9	21·0	21·9	22·7
218	3·5	7 49	23·3	23·6	23·3	22·8	22·3	21·8	21·7	21·9	22·5	23·3	24·2	25·1	25·8
219	3·8	7 52	18·2	18·4	18·1	17·4	16·7	16·1	15·8	16·0	16·5	17·4	18·5	19·5	20·2
220	3·6	7 56	51·4	51·6	51·1	50·2	49·2	48·3	47·9	47·9	48·5	49·5	50·8	52·0	52·8
221	2·3	8 03	40·2	40·5	40·2	39·6	38·8	38·3	38·0	38·1	38·6	39·4	40·5	41·5	42·3
222	2·9	8 07	38·3	38·6	38·5	38·0	37·4	37·0	36·8	37·0	37·5	38·2	39·2	40·1	40·9
223	1·9	8 09	36·9	37·2	36·8	36·1	35·2	34·5	34·1	34·1	34·6	35·6	36·7	37·8	38·7
224	3·8	8 16	37·6	38·1	38·0	37·6	37·2	36·9	36·8	37·0	37·5	38·3	39·2	40·1	41·0
225	4·4	8 18	38·6	38·9	38·7	38·2	37·5	36·9	36·7	36·7	37·2	38·0	39·0	40·0	40·8
226	4·4	8 22	58·6	59·2	59·2	58·7	58·1	57·6	57·5	57·7	58·4	59·3	60·6	61·8	62·9
227	4·3	8 20	41·1	41·2	39·7	37·0	33·8	30·8	28·6	27·8	28·6	30·8	33·8	36·7	38·6
228	1·7	8 22	35·3	35·6	35·1	34·1	32·9	31·8	31·0	30·9	31·4	32·5	33·9	35·3	36·4
229	3·9	8 25	45·8	46·3	46·2	45·9	45·4	45·1	45·0	45·1	45·6	46·3	47·2	48·1	48·9
230	3·6	8 25	48·3	48·5	47·9	46·6	45·0	43·5	42·5	42·2	42·7	44·0	45·7	47·4	48·7
231	3·5	8 30	26·6	27·4	27·4	26·6	25·5	24·7	24·4	24·6	25·4	26·8	28·5	30·3	31·9
232	4·2	8 37	45·9	46·3	46·3	46·0	45·6	45·2	45·1	45·3	45·7	46·4	47·3	48·3	49·1
233	4·1	8 37	43·8	44·2	44·0	43·4	42·7	42·0	41·6	41·6	42·0	42·7	43·8	44·9	45·8
234	3·7	8 40	22·4	22·8	22·5	21·8	20·8	19·9	19·4	19·2	19·6	20·5	21·7	23·0	24·0
235	4·1	8 40	42·6	43·0	42·8	42·2	41·4	40·7	40·2	40·1	40·5	41·3	42·4	43·6	44·5
236	3·7	8 43	41·0	41·4	41·3	40·8	40·2	39·7	39·4	39·4	39·8	40·5	41·4	42·5	43·3
237	4·2	8 44	48·0	48·6	48·6	48·3	47·8	47·5	47·3	47·5	47·9	48·7	49·6	50·6	51·5
238	2·0	8 44	47·0	47·4	47·1	46·4	45·4	44·4	43·8	43·6	44·0	44·9	46·1	47·5	48·5
239	4·2	8 46	49·2	49·8	49·8	49·5	49·0	48·6	48·5	48·6	49·1	49·9	50·9	52·0	52·9
240	3·5	8 46	53·0	53·6	53·6	53·3	52·8	52·5	52·4	52·5	52·9	53·6	54·5	55·4	56·3
241	4·2	8 50	37·5	37·9	37·9	37·5	36·9	36·5	36·2	36·2	36·5	37·2	38·1	39·1	40·0
242	3·3	8 55	30·0	30·6	30·6	30·3	29·9	29·6	29·4	29·5	29·9	30·6	31·4	32·4	33·3
243	4·0	8 55	07·4	07·9	07·6	06·7	05·5	04·3	03·4	03·1	03·4	04·4	05·8	07·3	08·5
244	4·3	8 58	35·8	36·4	36·5	36·2	35·8	35·4	35·3	35·4	35·8	36·4	37·3	38·3	39·2
245	3·1	8 59	20·9	21·7	21·8	21·3	20·7	20·1	19·8	19·9	20·4	21·4	22·6	24·0	25·2
246	4·1	9 00	46·3	47·1	47·2	46·8	46·2	45·7	45·4	45·5	46·0	46·9	48·0	49·3	50·4
247	4·4	9 00	10·6	11·1	11·0	10·5	09·9	09·2	08·8	08·7	09·0	09·7	10·7	11·8	12·8
248	3·7	9 03	45·9	46·8	46·9	46·4	45·8	45·2	44·9	45·0	45·5	46·4	47·6	49·0	50·3
249	4·2	9 02	31·1	31·7	31·3	30·2	28·7	27·2	26·0	25·5	25·7	26·8	28·5	30·3	31·7
250	3·7	9 04	14·4	14·9	14·8	14·2	13·4	12·7	12·2	12·0	12·3	13·0	14·1	15·3	16·3

The figures given refer to the beginning of the month, and should be interpolated to the actual date by means of the table on page 73.

No.		Name	Dec	Jan.	F.	M.	Apr.	M.	J.	July	A.	S.	Oct.	N.	D.	Jan.
			° ′	″	″	″	″	″	″	″	″	″	″	″	″	″
201	π	Puppis	S 37 05	61	70	77	79	78	73	65	57	49	46	48	55	65
202	λ	Geminorum	N 16 32	12	11	11	12	13	14	15	15	16	15	13	10	08
203	δ	Volantis	S 67 57	34	45	53	57	56	51	42	32	23	19	20	27	38
204	δ	Geminorum	N 21 58	42	42	43	45	45	46	46	46	45	43	41	39	38
205	η	Canis Majoris	S 29 18	22	31	36	39	38	34	27	19	12	09	11	18	27
206	ι	Geminorum	N 27 47	37	39	41	42	43	43	42	40	39	37	35	33	32
207	β	Canis Minoris	N 8 17	07	05	04	04	05	06	09	11	12	12	10	06	02
208	ρ	Geminorum	N 31 46	48	50	53	55	55	54	53	51	49	46	44	43	43
209	σ	Puppis	S 43 18	15	25	32	36	35	31	22	13	05	02	03	10	20
210		Castor (α Gem)	N 31 52	61	63	65	67	68	67	66	64	61	59	56	55	55
211	υ	Geminorum	N 26 53	27	28	30	32	33	32	32	31	29	27	24	22	21
212		Procyon (α CMi)	N 5 13	12	09	07	07	08	10	13	15	17	16	14	10	05
213	α	Monocerotis	S 9 33	19	25	28	30	29	26	21	16	12	11	13	18	25
214	ζ	Volantis	S 72 36	31	42	51	56	57	53	44	34	25	20	20	26	37
215	κ	Geminorum	N 24 23	34	35	36	38	39	39	39	38	36	34	32	29	28
216		Pollux (β Gem)	N 28 01	15	16	19	20	21	21	20	19	17	14	12	10	09
217	c	Puppis	S 37 58	19	29	36	40	40	36	28	20	12	08	10	16	26
218	ξ	Puppis	S 24 51	49	57	63	66	66	62	56	48	42	40	41	47	56
219		BS 3080 (Puppis)	S 40 34	45	55	63	67	67	63	56	47	39	35	36	42	52
220	χ	Carinae	S 52 58	68	79	87	92	93	89	81	72	63	58	59	65	75
221	ζ	Puppis	S 40 00	25	35	42	47	47	44	37	28	20	16	17	23	33
222	ρ	Puppis	S 24 18	31	39	45	48	48	45	39	32	26	24	25	31	39
223	γ²	Velorum	S 47 20	25	35	44	49	50	46	39	30	22	17	17	23	33
224	β	Cancri	N 9 10	46	44	43	43	44	45	47	49	50	49	46	42	37
225		BS 3270 (Puppis)	S 36 39	48	58	66	70	71	68	61	53	46	42	42	48	57
226	31	Lyncis	N 43 10	50	54	58	62	64	63	60	56	51	46	42	40	41
227	θ	Chamaeleontis	S 77 29	15	26	36	43	45	43	36	27	17	11	09	14	24
228	ε	Carinae	S 59 30	46	58	67	74	76	73	66	57	47	41	41	46	56
229		BS 3314 (Hydrae)	S 3 54	44	49	52	53	53	51	47	43	41	40	42	47	53
230	β	Volantis	S 66 08	25	36	46	53	56	53	46	37	27	21	20	25	35
231	o	Ursae Majoris	N 60 42	35	42	49	54	56	54	49	42	35	28	23	22	25
232	δ	Hydrae	N 5 41	50	46	45	44	45	47	49	51	52	52	49	44	39
233		BS 3426 (Velorum)	S 42 59	36	46	55	61	62	60	54	45	37	32	32	37	46
234	o	Velorum	S 52 55	33	44	53	60	62	60	54	45	36	30	29	34	43
235		BS 3445 (Velorum)	S 46 39	10	21	30	36	38	36	29	21	12	07	06	11	21
236	α	Pyxidis	S 33 11	28	38	45	50	51	49	43	36	29	25	25	30	39
237	δ	Cancri	N 18 08	48	47	47	49	50	51	52	52	51	48	45	40	37
238	δ	Velorum	S 54 42	43	55	64	71	74	72	65	56	47	41	40	45	54
239	ι	Cancri	N 28 44	67	67	70	72	74	75	74	72	69	65	61	57	55
240	ε	Hydrae	N 6 24	43	39	37	37	38	40	42	44	45	44	41	36	31
241	γ	Pyxidis	S 27 42	54	63	70	74	75	73	68	61	55	51	52	57	65
242	ζ	Hydrae	N 5 56	18	14	13	13	14	15	17	19	20	19	16	11	06
243		BS 3571 (Carinae)	S 60 38	54	65	76	83	87	85	79	70	61	54	52	57	66
244	α	Cancri	N 11 50	60	58	57	58	59	60	62	63	63	61	57	53	48
245	ι	Ursae Majoris	N 48 01	56	60	65	70	73	73	70	65	58	52	46	43	44
246		BS 3579 (Lyncis)	N 41 46	24	27	32	36	39	39	37	32	27	21	16	13	12
247		BS 3591 (Velorum)	S 41 15	30	40	49	55	58	56	50	42	34	29	29	33	42
248	κ	Ursae Majoris	N 47 08	49	53	58	63	66	66	63	58	52	45	40	37	37
249	α	Volantis	S 66 23	59	71	81	89	93	93	87	78	69	61	59	63	72
250		BS 3614 (Velorum)	S 47 06	08	19	28	35	38	36	31	22	14	08	07	11	20

No.	Mag.	RA	Jan.	Feb.	Mar.	Apr.	May	June	July	Aug.	Sept.	Oct.	Nov.	Dec.	Jan.
		h m	s	s	s	s	s	s	s	s	s	s	s	s	s
251	2·2	9 08	05·0	05·5	05·4	04·9	04·2	03·5	03·1	02·9	03·2	03·9	04·9	06·1	07·1
252	3·6	9 10	62·7	63·3	63·1	62·3	61·2	60·1	59·3	58·9	59·1	60·0	61·3	62·8	64·1
253	1·8	9 13	16·1	16·8	16·3	15·1	13·4	11·6	10·2	09·4	09·6	10·7	12·6	14·7	16·3
254	3·8	9 14	28·1	28·7	28·8	28·6	28·2	27·8	27·6	27·7	28·0	28·6	29·5	30·4	31·3
255	2·2	9 17	10·1	10·7	10·6	09·8	08·8	07·7	06·8	06·4	06·6	07·4	08·7	10·2	11·5
256	3·8	9 18	58·1	58·9	59·1	58·8	58·3	57·8	57·6	57·6	58·0	58·7	59·8	60·9	62·1
257	3·3	9 21	10·6	11·4	11·5	11·3	10·8	10·3	10·1	10·1	10·5	11·2	12·2	13·3	14·4
258	2·6	9 22	11·7	12·3	12·2	11·6	10·7	09·8	09·0	08·7	08·9	09·6	10·8	12·2	13·4
259	2·2	9 27	41·2	41·8	41·9	41·7	41·3	40·9	40·7	40·7	40·9	41·5	42·3	43·2	44·2
260	3·7	9 31	41·7	43·0	43·2	42·7	41·7	40·7	40·0	39·8	40·3	41·4	43·1	45·0	46·8
261	3·6	9 30	47·2	47·8	47·9	47·5	46·9	46·3	45·8	45·6	45·8	46·4	47·3	48·5	49·5
262	3·0	9 31	18·2	18·9	18·8	18·2	17·2	16·2	15·4	15·0	15·1	15·8	17·1	18·5	19·8
263	3·3	9 32	59·6	60·6	60·8	60·5	59·7	59·1	58·6	58·6	59·0	59·8	61·0	62·5	63·9
264	4·1	9 39	57·4	58·1	58·3	58·1	57·7	57·3	57·1	57·1	57·3	57·8	58·6	59·6	60·5
265	3·8	9 41	15·3	16·0	16·2	16·0	15·7	15·3	15·1	15·1	15·3	15·9	16·7	17·7	18·6
266	3·1	9 45	57·7	58·5	58·7	58·6	58·2	57·8	57·5	57·5	57·8	58·3	59·2	60·2	61·3
267	4–5	9 45	19·6	20·4	20·4	19·7	18·6	17·3	16·3	15·6	15·7	16·4	17·8	19·4	20·9
268	3·1	9 47	10·9	11·8	11·7	11·0	09·8	08·4	07·2	06·4	06·4	07·2	08·7	10·4	12·1
269	3·9	9 51	08·2	09·4	09·7	09·4	08·5	07·7	07·0	06·8	07·1	08·0	09·4	11·1	12·8
270	4·1	9 52	52·5	53·3	53·5	53·4	53·0	52·6	52·3	52·3	52·5	53·1	53·9	55·0	56·0
271	3·7	9 56	56·8	57·6	57·7	57·2	56·5	55·6	54·8	54·3	54·3	54·9	56·0	57·4	58·7
272	3·6	10 07	26·3	27·0	27·3	27·2	26·9	26·5	26·3	26·2	26·4	26·8	27·6	28·6	29·6
273 d	1·3	10 08	28·5	29·2	29·5	29·4	29·1	28·7	28·5	28·4	28·5	29·0	29·8	30·8	31·8
274	3·8	10 10	41·0	41·7	41·9	41·8	41·5	41·1	40·8	40·7	40·8	41·3	42·0	43·0	43·9
275	3·6	10 13	48·9	50·1	50·3	49·5	48·1	46·4	44·8	43·6	43·4	44·1	45·7	47·8	49·8
276	4·1	10 14	49·4	50·2	50·4	50·2	49·7	49·1	48·6	48·2	48·2	48·7	49·6	50·7	51·9
277	3·6	10 16	47·8	48·6	49·0	48·9	48·5	48·2	47·9	47·8	47·9	48·4	49·2	50·2	51·3
278	3·5	10 17	12·9	13·9	14·3	14·1	13·7	13·1	12·7	12·5	12·7	13·3	14·2	15·4	16·7
279	3·4	10 17	10·0	10·9	11·1	10·7	09·8	08·7	07·6	06·9	06·7	07·3	08·5	10·1	11·7
280	2·6	10 20	04·7	05·5	05·8	05·8	05·4	05·1	04·8	04·7	04·8	05·3	06·1	07·1	08·1
281	3·2	10 22	26·7	27·7	28·1	28·0	27·5	27·0	26·6	26·4	26·6	27·1	28·0	29·2	30·5
282	4·1	10 24	28·3	29·9	30·1	29·3	27·6	25·4	23·3	21·7	21·2	21·9	23·8	26·4	28·9
283	4·1	10 26	11·0	11·8	12·1	12·0	11·7	11·3	11·0	10·8	10·9	11·3	12·0	12·9	13·9
284	4·4	10 27	14·6	15·3	15·6	15·5	15·1	14·7	14·2	14·0	14·0	14·4	15·2	16·2	17·3
285	4·4	10 27	59·7	60·6	61·0	61·0	60·6	60·1	59·7	59·6	59·7	60·2	61·0	62·1	63·3
286	4·1	10 27	57·8	58·8	59·1	58·8	58·0	57·0	56·0	55·4	55·2	55·7	56·8	58·3	59·8
287	3·6	10 32	06·5	07·6	07·9	07·5	06·7	05·6	04·5	03·7	03·5	04·0	05·1	06·7	08·3
288	3·8	10 32	54·6	55·5	55·8	55·8	55·5	55·2	54·9	54·7	54·8	55·2	55·9	56·9	57·9
289	4·1	10 35	32·7	34·8	35·2	34·2	31·9	28·8	25·7	23·3	22·3	23·1	25·6	29·0	32·4
290	4·4	10 39	23·6	24·6	24·9	24·7	24·0	23·2	22·3	21·7	21·5	21·9	22·9	24·3	25·7
291	3·0	10 42	62·4	63·7	64·1	63·7	62·8	61·6	60·4	59·5	59·1	59·6	60·8	62·5	64·2
292	2·8	10 46	51·5	52·4	52·8	52·7	52·2	51·5	50·8	50·3	50·1	50·5	51·4	52·6	53·9
293	3·3	10 49	43·1	43·9	44·3	44·2	44·0	43·6	43·3	43·1	43·1	43·4	44·1	45·0	46·0
294	3·9	10 53	25·0	26·0	26·5	26·5	26·2	25·8	25·4	25·1	25·2	25·5	26·3	27·4	28·5
295	3·9	10 53	34·8	36·0	36·4	36·2	35·6	34·7	33·7	32·9	32·6	33·0	34·0	35·4	37·0
296	4·2	10 59	51·9	52·8	53·2	53·2	52·9	52·6	52·2	52·0	52·0	52·2	52·9	53·8	54·8
297	2·4	11 01	57·5	58·9	59·6	59·6	59·0	58·3	57·5	57·0	57·0	57·4	58·4	59·8	61·4
298	1·9	11 03	50·9	52·5	53·3	53·3	52·6	51·7	50·8	50·2	50·0	50·5	51·6	53·2	55·0
299	4·0	11 08	40·7	41·9	42·4	42·3	41·7	40·9	39·9	39·1	38·7	39·0	39·9	41·4	43·0
300	3·1	11 09	46·1	47·3	47·9	47·9	47·6	47·1	46·5	46·2	46·1	46·5	47·2	48·4	49·7

The figures given refer to the beginning of the month, and should be interpolated to the actual date by means of the table on page 73.

No.		Name	Dec	Jan.	F.	M.	Apr.	M.	J.	July	A.	S.	Oct.	N.	D.	Jan.
			° ′	″	″	″	″	″	″	″	″	″	″	″	″	″
251	λ	Velorum	S 43 26	14	25	34	40	43	41	36	28	20	14	13	18	27
252	a	Carinae	S 58 58	16	27	38	45	50	49	43	35	25	19	16	20	29
253	β	Carinae	S 69 43	15	26	37	45	50	50	45	36	27	19	16	19	28
254	θ	Hydrae	N 2 18	24	19	17	16	17	18	21	23	24	24	21	16	10
255	ι	Carinae	S 59 16	45	57	67	75	80	79	74	65	56	49	47	50	59
256	38	Lyncis	N 36 47	34	36	39	43	46	47	46	43	38	32	27	22	20
257	α	Lyncis	N 34 22	59	60	63	67	70	70	69	67	62	57	51	47	45
258	κ	Velorum	S 55 00	54	65	75	83	87	87	82	73	64	58	55	59	68
259	α	Hydrae	S 8 39	57	63	68	70	70	69	65	61	58	57	59	64	71
260	23	Ursae Majoris	N 63 02	63	68	75	82	86	86	82	75	67	58	51	47	48
261	ψ	Velorum	S 40 28	18	28	37	44	47	46	42	34	27	21	20	24	32
262	N	Velorum	S 57 02	19	30	41	49	54	54	49	41	32	25	22	25	33
263	θ	Ursae Majoris	N 51 39	58	62	68	74	78	78	75	70	62	55	48	44	43
264	ι	Hydrae	S 1 09	03	08	12	13	13	11	08	06	04	04	07	12	18
265	ο	Leonis	N 9 52	61	57	56	56	58	59	61	62	62	60	56	51	46
266	ε	Leonis	N 23 45	52	51	52	54	57	59	59	58	55	51	46	41	36
267	l	Carinae	S 62 30	43	54	65	74	80	81	77	69	60	52	48	51	59
268	υ	Carinae	S 65 04	33	44	55	64	70	71	67	60	50	42	39	41	48
269	υ	Ursae Majoris	N 59 01	37	41	48	55	59	60	57	51	42	33	26	21	21
270	μ	Leonis	N 25 59	48	47	49	52	55	56	56	55	52	47	42	36	32
271	φ	Velorum	S 54 34	21	31	42	50	56	57	53	46	37	30	26	29	37
272	η	Leonis	N 16 44	70	67	67	69	71	73	74	74	72	69	64	59	53
273	*Regulus* (α Leo)		N 11 57	28	24	23	24	25	27	29	29	29	26	22	16	11
274	λ	Hydrae	S 12 21	42	50	55	58	59	58	55	51	48	46	48	52	59
275	ω	Carinae	S 70 02	31	42	53	63	70	73	70	63	54	45	40	41	48
276	BS 4023 (Velorum)		S 42 07	39	49	58	66	71	71	68	61	54	48	46	48	56
277	ζ	Leonis	N 23 24	24	22	23	26	29	31	31	31	28	23	18	12	07
278	λ	Ursae Majoris	N 42 53	69	70	74	80	84	86	85	82	76	68	61	55	52
279	BS 4050 (Carinae)		S 61 20	12	23	34	43	50	52	49	42	33	25	21	22	29
280*	γ¹	Leonis	N 19 49	52	49	49	52	54	56	57	57	55	51	45	39	34
281	μ	Ursae Majoris	N 41 28	75	76	80	86	90	92	92	88	82	75	67	61	58
282	I	Carinae	S 74 02	08	18	30	40	48	51	49	43	33	25	19	19	25
283	μ	Hydrae	S 16 50	38	46	52	56	58	57	54	50	46	44	44	48	55
284	α	Antliae	S 31 04	27	36	44	50	54	54	51	46	40	35	34	37	44
285	β	Leonis Minoris	N 36 41	43	43	47	51	56	58	58	55	50	43	36	30	26
286	BS 4114 (Carinae)		S 58 44	38	49	60	69	76	78	76	69	60	53	48	49	56
287	BS 4140 (Carinae)		S 61 41	23	34	45	54	62	64	62	55	47	39	34	34	41
288	ρ	Leonis	N 9 17	48	44	42	43	44	46	48	49	49	46	42	37	31
289	γ	Chamaeleontis	S 78 36	42	52	63	74	82	86	85	79	70	61	55	54	60
290	BS 4180 (Velorum)		S 55 36	29	39	50	59	66	68	66	60	52	44	40	41	47
291	θ	Carinae	S 64 23	56	66	77	87	95	98	96	90	82	73	68	68	74
292	μ	Velorum	S 49 25	31	41	51	60	66	68	66	60	52	45	41	43	49
293	ν	Hydrae	S 16 12	05	12	19	23	25	24	22	18	14	12	12	16	23
294	46	Leonis Minoris	N 34 11	69	68	71	75	80	82	83	81	76	70	62	55	50
295	u	Carinae	S 58 51	29	39	50	60	67	71	69	63	55	47	42	42	48
296	α	Crateris	S 18 18	23	31	37	42	44	44	42	38	34	31	31	35	42
297	β	Ursae Majoris	N 56 21	68	69	75	82	89	92	91	86	79	69	60	52	49
298	*Dubhe* (α UMa)		N 61 43	73	75	82	90	96	99	98	93	84	74	64	57	54
299	BS 4337 (Carinae)		S 58 58	47	57	67	77	85	89	88	83	75	67	61	61	66
300	ψ	Ursae Majoris	N 44 28	67	67	71	77	83	86	86	83	77	69	60	53	48

* No., mag., dist. and p.a. of companion star:　**280**, 3·8, 4″, 125°

No.	Mag.	RA	Jan.	Feb.	Mar.	Apr.	May	June	July	Aug.	Sept.	Oct.	Nov.	Dec.	Jan.
		h m	s	s	s	s	s	s	s	s	s	s	s	s	s
301	2·6	11 14	12·3	13·3	13·8	13·8	13·6	13·3	13·0	12·8	12·7	13·0	13·6	14·6	15·6
302	3·4	11 14	20·1	21·0	21·5	21·6	21·4	21·1	20·8	20·6	20·6	20·8	21·4	22·3	23·4
303	3·9	11 18	16·8	17·8	18·4	18·5	18·2	17·9	17·5	17·2	17·1	17·4	18·1	19·0	20·2
304	3·7	11 18	34·7	35·7	36·3	36·4	36·1	35·7	35·3	35·0	35·0	35·3	35·9	36·9	38·1
305	3·8	11 19	25·9	26·8	27·3	27·3	27·2	26·9	26·5	26·3	26·2	26·4	27·0	27·9	28·9
306	4·1	11 21	13·8	14·7	15·2	15·3	15·1	14·8	14·5	14·3	14·2	14·5	15·1	15·9	17·0
307	4·3	11 21	05·8	06·9	07·5	07·5	07·1	06·4	05·6	04·9	04·5	04·7	05·5	06·8	08·3
308	4·1	11 24	58·4	59·3	59·7	59·8	59·7	59·4	59·0	58·7	58·6	58·8	59·4	60·3	61·4
309	4·1	11 31	31·2	33·3	34·5	34·7	33·9	32·6	31·2	30·1	29·6	29·9	31·1	33·0	35·3
310	3·7	11 33	05·5	06·4	07·0	07·1	06·9	06·5	06·1	05·7	05·5	05·7	06·3	07·3	08·4
311	3·3	11 35	52·3	53·7	54·5	54·5	54·0	53·1	52·0	51·0	50·4	50·5	51·4	52·9	54·7
312	3·8	11 45	41·9	43·6	44·5	44·6	44·1	43·0	41·7	40·4	39·6	39·7	40·6	42·3	44·3
313	4·2	11 45	57·0	57·9	58·5	58·7	58·6	58·3	58·0	57·7	57·6	57·8	58·3	59·1	60·2
314	3·8	11 46	08·7	10·0	10·8	11·0	10·7	10·2	09·6	09·1	08·9	09·0	09·7	10·8	12·1
315	4·2	11 46	36·3	37·7	38·5	38·6	38·3	37·4	36·4	35·5	34·9	34·9	35·7	37·2	38·9
316	2·2	11 49	08·9	09·9	10·5	10·7	10·6	10·3	10·0	09·7	09·6	09·7	10·2	11·1	12·1
317	3·8	11 50	47·2	48·2	48·7	48·9	48·8	48·6	48·3	48·0	47·9	48·1	48·6	49·4	50·4
318	2·5	11 53	55·5	56·9	57·8	58·0	57·8	57·1	56·4	55·8	55·5	55·6	56·3	57·5	58·9
319	4·2	12 05	17·8	18·8	19·4	19·6	19·6	19·3	19·0	18·7	18·6	18·7	19·1	19·9	21·0
320	2·9	12 08	27·0	28·2	29·0	29·3	29·1	28·6	28·0	27·3	26·8	26·8	27·4	28·5	29·9
321	3·2	12 10	12·8	13·8	14·4	14·6	14·6	14·4	14·0	13·7	13·4	13·5	14·0	14·8	15·9
322	3·1	12 15	14·3	15·7	16·6	17·0	16·8	16·2	15·3	14·4	13·7	13·7	14·3	15·6	17·2
323	3·4	12 15	30·7	32·3	33·3	33·7	33·4	32·8	32·0	31·2	30·7	30·7	31·3	32·5	34·0
324	2·8	12 15	53·6	54·6	55·2	55·5	55·5	55·3	55·0	54·6	54·4	54·4	54·9	55·7	56·7
325	4·4	12 18	27·3	30·8	33·0	33·6	32·8	30·6	27·6	24·5	22·2	21·5	22·8	25·9	29·9
326	4·0	12 19	59·6	60·5	61·2	61·4	61·4	61·3	61·0	60·7	60·5	60·5	60·9	61·7	62·7
327	3·6	12 21	27·2	28·8	29·7	30·1	30·0	29·3	28·4	27·4	26·7	26·6	27·2	28·5	30·2
328 d	1·6	12 26	41·6	43·3	44·3	44·8	44·6	44·0	42·9	41·8	41·0	40·8	41·4	42·8	44·6
329	4·2	12 28	07·9	09·2	10·0	10·4	10·3	10·0	09·3	08·6	08·1	08·0	08·5	09·6	11·0
330	3·1	12 29	57·1	58·1	58·7	59·0	59·1	58·9	58·6	58·2	58·0	58·0	58·4	59·1	60·2
331	1·6	12 31	15·6	17·0	18·0	18·4	18·3	17·9	17·1	16·2	15·5	15·4	15·9	17·1	18·7
332	4·0	12 32	34·1	36·4	37·9	38·6	38·3	37·2	35·6	33·8	32·4	32·0	32·7	34·6	37·2
333	3·9	12 33	33·2	35·5	37·1	37·7	37·4	36·3	34·8	33·4	32·4	32·0	32·6	34·2	36·3
334	4·3	12 33	49·3	50·5	51·3	51·7	51·6	51·3	50·8	50·3	49·9	49·9	50·3	51·2	52·3
335	2·8	12 34	28·5	29·5	30·2	30·5	30·6	30·4	30·1	29·7	29·4	29·4	29·8	30·6	31·6
336	2·9	12 37	17·0	19·0	20·4	21·0	20·9	20·0	18·7	17·2	16·0	15·6	16·3	17·9	20·1
337	2·4	12 41	36·5	37·8	38·7	39·1	39·1	38·8	38·2	37·5	37·0	36·9	37·3	38·3	39·6
338	2·9	12 41	44·6	45·6	46·3	46·6	46·7	46·5	46·3	45·9	45·7	45·7	46·0	46·7	47·7
339	3·3	12 46	23·1	25·1	26·5	27·2	27·1	26·4	25·1	23·7	22·6	22·2	22·7	24·3	26·4
340	1·5	12 47	49·0	50·6	51·7	52·2	52·2	51·8	50·9	50·0	49·2	48·9	49·4	50·6	52·2
341	4·3	12 53	31·6	32·8	33·6	34·1	34·2	34·0	33·5	33·0	32·5	32·4	32·8	33·6	34·8
342	1·7	12 54	05·9	07·5	08·6	09·2	09·1	08·6	07·9	07·1	06·4	06·2	06·5	07·5	08·9
343	3·7	12 55	41·1	42·1	42·8	43·2	43·3	43·2	42·9	42·6	42·3	42·3	42·6	43·2	44·2
344	2·9	12 56	06·2	07·4	08·2	08·7	08·7	08·5	08·0	07·5	07·1	07·0	07·3	08·1	09·2
345	3·6	13 02	22·7	25·1	26·9	27·8	27·9	27·1	25·6	23·9	22·4	21·8	22·3	24·0	26·4
346	2·9	13 02	15·4	16·4	17·2	17·6	17·7	17·6	17·3	16·9	16·6	16·6	16·9	17·5	18·5
347	4·4	13 07	00·2	01·6	02·6	03·1	03·3	03·1	02·5	01·8	01·2	01·0	01·3	02·2	03·6
348	4·5	13 10	02·0	03·0	03·7	04·1	04·3	04·2	04·0	03·6	03·3	03·2	03·5	04·2	05·1
349	4·3	13 11	56·8	57·9	58·7	59·2	59·3	59·1	58·8	58·3	58·0	57·8	58·1	58·7	59·7
350	3·3	13 19	00·5	01·5	02·3	02·8	03·0	02·9	02·7	02·3	01·9	01·8	02·0	02·7	03·7

The figures given refer to the beginning of the month, and should be interpolated to the actual date by means of the table on page 73.

No.		Name	Dec	Jan.	F.	M.	Apr.	M.	J.	July	A.	S.	Oct.	N.	D.	Jan.
			° ′	″	″	″	″	″	″	″	″	″	″	″	″	″
301	δ	Leonis	N 20 30	44	41	41	43	46	49	50	50	48	43	37	31	25
302	θ	Leonis	N 15 24	67	63	62	63	66	68	70	70	69	65	60	54	47
303	ξ	Ursae Majoris	N 31 30	59	57	59	63	68	71	72	70	66	60	52	45	39
304	ν	Ursae Majoris	N 33 04	55	53	55	59	64	67	68	67	62	56	48	41	35
305	δ	Crateris	S 14 47	12	19	25	29	31	31	29	25	22	20	21	24	31
306	σ	Leonis	N 6 00	69	64	61	61	62	64	66	67	68	66	62	56	50
307	π	Centauri	S 54 29	45	55	65	75	82	86	85	81	73	65	60	60	65
308	γ	Crateris	S 17 41	31	38	45	49	52	52	50	47	43	40	41	44	50
309	λ	Draconis	N 69 18	60	62	68	77	84	88	87	82	73	62	51	43	40
310	ξ	Hydrae	S 31 51	51	60	68	75	80	81	80	76	71	66	64	65	71
311	λ	Centauri	S 63 01	27	36	47	57	66	71	71	67	59	51	45	43	47
312	λ	Muscae	S 66 43	59	67	78	88	98	03	104	101	93	85	78	75	79
313	ν	Virginis	N 6 30	68	63	60	60	61	63	65	66	66	65	60	54	48
314	χ	Ursae Majoris	N 47 45	57	56	60	66	73	77	78	75	69	60	51	42	36
315		BS 4522 (Centauri)	S 61 10	58	67	77	88	97	02	102	98	91	83	77	75	79
316		*Denebola (β Leo)*	N 14 33	39	35	33	34	37	40	42	42	41	38	32	26	19
317	β	Virginis	N 1 44	77	71	68	66	67	69	71	72	73	72	68	63	56
318	γ	Ursae Majoris	N 53 40	50	50	54	61	69	73	74	71	64	55	45	36	30
319	o	Virginis	N 8 43	21	15	13	13	15	17	19	21	20	18	13	07	00
320	δ	Centauri	S 50 43	39	47	57	66	74	78	79	76	69	62	57	56	60
321	ε	Corvi	S 22 37	38	45	52	58	61	62	61	58	55	51	50	53	58
322	δ	Crucis	S 58 45	12	20	30	40	49	55	56	53	47	39	33	30	33
323	δ	Ursae Majoris	N 57 00	66	65	69	77	84	90	91	89	82	72	61	52	45
324	γ	Corvi	S 17 32	59	67	73	77	80	81	79	77	74	71	71	74	80
325	β	Chamaeleontis	S 79 18	57	64	74	85	96	04	107	105	98	90	81	77	78
326	η	Virginis	S 0 40	35	41	45	47	46	45	43	41	40	41	44	49	56
327	ε	Crucis	S 60 24	21	29	38	48	58	64	66	63	56	49	42	39	42
328*	α	Crucis	S 63 06	12	19	29	39	49	55	58	55	49	41	34	30	33
329	σ	Centauri	S 50 14	09	16	25	34	42	47	49	46	40	33	28	26	29
330	δ	Corvi	S 16 31	24	31	37	42	44	45	44	42	39	37	37	39	45
331	γ	Crucis	S 57 07	04	12	21	31	40	46	48	45	39	32	26	23	25
332	γ	Muscae	S 72 08	12	19	29	40	50	58	61	59	53	44	36	32	33
333	κ	Draconis	N 69 46	25	24	29	37	46	52	54	50	43	32	20	10	04
334	β	Canum Venat.	N 41 20	40	37	39	44	51	56	58	57	53	45	36	26	19
335	β	Corvi	S 23 24	14	22	28	34	38	39	39	36	33	29	28	30	35
336	α	Muscae	S 69 08	22	29	38	49	59	67	70	68	62	54	46	42	43
337	γ	Centauri	S 48 57	53	60	69	77	85	90	92	90	84	78	72	70	73
338†	γ	Virginis	S 1 27	31	38	42	43	43	42	40	38	37	38	41	45	52
339	β	Muscae	S 68 06	44	50	59	70	80	87	91	89	84	75	68	63	64
340	β	Crucis	S 59 41	35	42	51	61	70	77	79	77	72	64	58	54	56
341		BS 4889 (Centauri)	S 40 11	04	11	19	27	34	38	39	37	32	27	23	22	25
342	ε	Ursae Majoris	N 55 56	45	43	46	53	61	67	70	68	63	53	42	32	24
343	δ	Virginis	N 3 22	76	69	66	65	66	68	70	72	72	71	67	62	55
344	α	Canum Venat.	N 38 17	80	76	77	82	88	94	97	97	93	86	77	67	59
345	δ	Muscae	S 71 33	09	15	23	34	44	53	57	56	51	43	35	30	30
346	ε	Virginis	N 10 56	55	50	47	48	50	53	55	57	56	54	49	42	35
347	ξ²	Centauri	S 49 54	40	46	54	63	71	77	79	78	73	67	61	58	60
348	θ	Virginis	S 5 32	52	58	63	65	66	65	63	61	60	60	62	66	72
349	β	Comae Ber.	N 27 51	61	56	55	58	64	69	72	73	71	65	58	49	41
350	γ	Hydrae	S 23 10	42	49	55	60	64	66	66	64	61	58	57	58	63

* No., mag., dist. and p.a. of companion star: **328**, 2·1, 4″, 113°

† **338**: position refers to midpoint of double star; mags., 3·6 and 3·7, separation, 1″

No.	Mag.	RA	Jan.	Feb.	Mar.	Apr.	May	June	July	Aug.	Sept.	Oct.	Nov.	Dec.	Jan.
		h m	s	s	s	s	s	s	s	s	s	s	s	s	s
351	2·9	13 20	41·1	42·3	43·2	43·7	43·9	43·8	43·5	43·0	42·5	42·3	42·6	43·3	44·4
352	2·4	13 23	59·0	60·6	61·8	62·4	62·5	62·2	61·5	60·6	59·9	59·5	59·7	60·5	61·8
353 d	1·2	13 25	16·6	17·6	18·4	18·8	19·0	19·0	18·8	18·4	18·1	17·9	18·2	18·8	19·8
354	4·0	13 31	08·1	09·3	10·2	10·8	11·0	11·0	10·6	10·1	09·6	09·4	09·6	10·3	11·5
355	3·4	13 34	46·3	47·3	48·1	48·6	48·8	48·8	48·6	48·3	47·9	47·8	47·9	48·5	49·4
356	2·6	13 39	59·0	60·5	61·6	62·4	62·7	62·6	62·1	61·3	60·6	60·2	60·4	61·2	62·6
357	4·4	13 45	46·4	47·6	48·5	49·1	49·4	49·4	49·1	48·7	48·2	47·9	48·1	48·8	49·8
358	1·9	13 47	35·7	37·1	38·2	38·9	39·1	38·9	38·4	37·6	36·9	36·5	36·6	37·2	38·3
359	3·5	13 49	35·7	37·0	38·0	38·7	39·0	39·0	38·7	38·1	37·6	37·3	37·4	38·1	39·3
360	3·3	13 49	42·4	43·7	44·7	45·4	45·8	45·8	45·4	44·9	44·3	44·0	44·1	44·8	46·0
361	2·8	13 54	45·3	46·4	47·2	47·8	48·1	48·1	47·8	47·4	47·0	46·8	46·9	47·4	48·3
362	3·1	13 55	38·0	39·3	40·4	41·2	41·6	41·6	41·2	40·6	39·9	39·5	39·7	40·4	41·6
363	4·3	14 01	43·4	44·4	45·2	45·8	46·1	46·1	46·0	45·6	45·2	45·0	45·1	45·6	46·5
364 d	0·9	14 03	55·5	57·2	58·7	59·7	60·2	60·2	59·6	58·7	57·7	57·1	57·1	58·0	59·5
365	3·6	14 04	24·8	26·7	28·3	29·4	29·7	29·3	28·3	27·1	25·8	25·0	24·8	25·4	26·8
366	3·5	14 06	27·4	28·5	29·4	30·1	30·4	30·5	30·3	29·9	29·5	29·2	29·3	29·8	30·8
367	2·3	14 06	46·2	47·4	48·4	49·0	49·4	49·5	49·3	48·8	48·2	47·9	48·0	48·6	49·7
368	4·3	14 12	58·5	59·5	60·4	61·0	61·3	61·4	61·3	61·0	60·6	60·3	60·4	60·9	61·8
369 d	0·2	14 15	43·6	44·6	45·5	46·1	46·4	46·5	46·3	45·9	45·4	45·1	45·1	45·6	46·4
370	4·2	14 16	05·6	06·6	07·4	08·0	08·4	08·5	08·4	08·0	07·6	07·4	07·4	07·9	08·8
371	4·3	14 16	26·0	27·2	28·4	29·1	29·4	29·4	28·9	28·3	27·6	27·1	27·0	27·5	28·5
372	4·4	14 20	25·5	27·1	28·4	29·4	30·0	30·1	29·7	28·9	28·0	27·4	27·4	28·1	29·5
373	4·2	14 20	38·8	40·0	41·0	41·7	42·2	42·3	42·1	41·6	41·1	40·7	40·8	41·3	42·4
374	4·1	14 25	14·2	15·5	16·8	17·6	18·0	17·9	17·4	16·6	15·8	15·1	15·0	15·4	16·4
375	4·4	14 27	28·8	31·7	34·5	36·4	36·9	36·2	34·4	32·0	29·5	27·6	26·8	27·4	29·3
376	3·8	14 31	53·3	54·4	55·4	56·0	56·4	56·5	56·2	55·8	55·3	54·9	54·8	55·2	56·0
377	3·0	14 32	07·8	09·0	10·0	10·7	11·1	11·2	10·9	10·3	09·7	09·2	09·1	09·5	10·4
378	2·6	14 35	35·9	37·2	38·3	39·1	39·6	39·8	39·6	39·1	38·5	38·1	38·0	38·6	39·6
379 d	0·1	14 39	41·7	43·5	45·0	46·2	46·8	46·9	46·4	45·5	44·4	43·6	43·5	44·2	45·6
380	3·9	14 41	13·0	14·0	14·9	15·5	15·9	16·1	15·9	15·6	15·1	14·8	14·8	15·1	15·9
381	2·9	14 42	01·4	02·8	04·0	04·9	05·5	05·7	05·5	04·9	04·2	03·7	03·6	04·2	05·3
382	3·4	14 42	36·8	38·9	40·7	42·1	42·9	43·1	42·6	41·5	40·3	39·4	39·1	39·9	41·5
383	3·9	14 43	08·2	09·2	10·1	10·7	11·2	11·3	11·2	10·9	10·5	10·2	10·2	10·6	11·4
384	4·1	14 43	44·7	45·9	46·9	47·7	48·2	48·4	48·3	47·9	47·3	46·9	46·9	47·4	48·3
385	2·7	14 45	02·8	03·8	04·8	05·5	05·9	06·0	05·8	05·4	04·9	04·5	04·4	04·7	05·5
386	3·8	14 48	00·4	04·6	08·3	11·3	13·1	13·3	11·9	09·2	06·2	03·8	03·0	04·3	07·5
387	3·8	14 46	19·2	20·2	21·1	21·8	22·2	22·4	22·3	22·0	21·5	21·2	21·2	21·6	22·3
388	2·9	14 50	57·5	58·5	59·4	60·1	60·6	60·8	60·7	60·4	60·0	59·6	59·6	60·0	60·8
389	2·2	14 50	39·3	41·9	44·4	46·3	47·1	46·7	45·2	43·1	40·7	38·8	37·8	38·1	39·7
390	2·8	14 58	37·4	38·7	39·8	40·8	41·4	41·7	41·5	41·1	40·4	39·9	39·8	40·3	41·3
391	3·3	14 59	15·1	16·4	17·5	18·4	19·1	19·4	19·2	18·8	18·1	17·7	17·5	18·0	19·0
392	3·6	15 01	59·4	60·6	61·6	62·5	63·0	63·1	62·8	62·3	61·6	61·0	60·8	61·1	61·9
393	3·4	15 04	09·2	10·2	11·2	12·0	12·5	12·8	12·8	12·4	11·9	11·5	11·5	11·8	12·7
394	4·1	15 12	01·7	03·1	04·4	05·4	06·2	06·6	06·4	05·9	05·2	04·6	04·4	04·8	05·9
395	3·5	15 12	22·9	24·4	25·7	26·8	27·6	28·0	27·9	27·3	26·5	25·8	25·6	26·0	27·2
396	3·5	15 15	33·1	34·2	35·2	36·0	36·5	36·7	36·5	36·1	35·5	35·0	34·7	35·0	35·7
397	4·2	15 17	37·0	38·7	40·2	41·6	42·5	42·9	42·7	42·0	41·0	40·2	39·9	40·3	41·6
398	2·7	15 17	04·9	05·9	06·8	07·5	08·1	08·4	08·3	08·1	07·6	07·2	07·1	07·4	08·2
399	3·1	15 19	01·5	03·8	05·9	07·7	09·0	09·6	09·2	08·0	06·5	05·3	04·7	05·3	07·0
400	3·4	15 21	27·7	28·9	30·1	31·0	31·7	32·1	32·1	31·7	31·0	30·5	30·3	30·7	31·6

The figures given refer to the beginning of the month, and should be interpolated to the actual date by means of the table on page 73.

No.		Name	Dec	Jan.	F.	M.	Apr.	M.	J.	July	A.	S.	Oct.	N.	D.	Jan.
			° ′	″	″	″	″	″	″	″	″	″	″	″	″	″
351	ι	Centauri	S 36 43	05	11	18	25	31	35	37	35	32	27	23	22	25
352 *	ζ	Ursae Majoris	N 54 54	43	39	41	48	56	63	67	66	61	53	42	31	23
353		Spica (α Vir)	S 11 10	09	16	21	24	26	26	25	23	21	20	21	24	29
354	d	Centauri	S 39 24	46	52	59	67	73	78	79	78	75	70	66	64	66
355	ζ	Virginis	S 0 36	17	23	27	29	28	27	25	23	22	23	26	31	37
356	ε	Centauri	S 53 28	14	19	27	35	44	50	54	54	50	44	38	34	34
357	I	Centauri	S 33 02	58	64	70	77	82	86	87	86	83	79	76	75	78
358	η	Ursae Majoris	N 49 17	62	57	58	64	72	79	83	84	81	73	63	52	43
359	ν	Centauri	S 41 41	34	39	46	53	60	65	67	67	63	59	54	52	53
360	μ	Centauri	S 42 28	43	48	55	62	69	74	77	77	73	68	64	61	63
361	η	Bootis	N 18 22	74	68	66	67	71	75	79	81	80	76	70	62	54
362	ζ	Centauri	S 47 17	34	39	46	53	61	67	70	70	67	62	56	53	53
363	τ	Virginis	N 1 31	70	63	59	58	59	61	63	65	66	65	62	57	50
364	β	Centauri	S 60 22	35	39	46	55	64	72	77	78	74	68	61	56	55
365	α	Draconis	N 64 21	47	42	43	50	59	67	72	73	68	60	49	37	28
366	π	Hydrae	S 26 41	18	23	29	34	38	41	42	41	39	36	34	34	37
367	θ	Centauri	S 36 22	31	36	43	49	55	59	61	61	58	54	51	49	51
368	κ	Virginis	S 10 16	51	57	62	65	66	66	65	63	62	61	62	65	70
369		Arcturus (α Boo)	N 19 09	77	71	68	69	73	78	82	84	83	79	73	65	57
370	ι	Virginis	S 6 00	30	36	41	43	43	42	41	39	38	38	40	44	49
371	λ	Bootis	N 46 04	36	30	30	35	42	50	55	57	54	48	38	27	18
372	ν	Centauri	S 56 23	24	28	34	42	50	58	63	64	61	56	49	44	43
373	ψ	Centauri	S 37 53	24	29	35	41	47	51	54	54	51	48	44	42	43
374	θ	Bootis	N 51 50	20	14	14	19	27	35	40	42	39	32	22	11	01
375	5	Ursae Minoris	N 75 40	61	56	57	64	73	82	87	88	84	75	64	52	43
376	ρ	Bootis	N 30 21	41	34	32	35	40	47	52	54	53	49	41	32	23
377	γ	Bootis	N 38 17	51	45	44	47	54	61	66	69	67	62	53	43	34
378	η	Centauri	S 42 09	43	47	53	59	65	71	74	75	73	69	64	61	61
379 *	α	Centauri	S 60 50	13	15	21	29	37	45	51	53	50	45	38	32	30
380	ζ	Bootis	N 13 42	70	63	60	61	64	68	72	74	75	72	67	60	52
381	α	Lupi	S 47 23	31	34	40	46	53	59	63	65	63	58	53	49	48
382	α	Circini	S 64 58	41	43	48	56	65	74	80	83	81	76	68	62	59
383	μ	Virginis	S 5 39	55	61	65	67	68	67	65	63	62	62	64	68	73
384		BS 5485 (Centauri)	S 35 10	42	46	51	57	62	66	69	69	67	64	61	59	60
385	ε	Bootis	N 27 03	53	46	44	46	51	57	62	65	64	60	53	44	36
386	α	Apodis	S 79 02	49	50	54	63	73	83	91	95	94	88	80	72	67
387	109	Virginis	N 1 52	67	61	57	56	57	60	62	64	65	64	61	56	50
388	α²	Librae	S 16 02	52	57	62	65	67	67	67	66	65	64	63	65	69
389	β	Ursae Minoris	N 74 08	38	32	33	39	48	57	63	65	62	54	43	31	21
390	β	Lupi	S 43 08	16	19	24	29	36	41	45	46	45	41	37	33	33
391	κ	Centauri	S 42 06	29	32	37	43	49	54	57	59	57	54	49	46	46
392	β	Bootis	N 40 22	51	43	42	45	52	59	65	69	68	63	55	44	34
393	σ	Librae	S 25 17	13	17	22	26	29	31	32	32	31	29	27	27	29
394	κ¹	Lupi	S 48 44	28	30	34	40	47	53	58	60	59	55	50	45	44
395	ζ	Lupi	S 52 06	08	10	14	20	27	34	39	42	41	37	31	26	24
396	δ	Bootis	N 33 18	21	14	11	13	19	26	32	36	36	32	25	15	05
397	β	Circini	S 58 48	14	15	19	25	33	41	47	50	49	45	39	33	30
398	β	Librae	S 9 23	19	25	29	31	31	31	30	28	27	27	28	30	35
399	γ	Trianguli Aust.	S 68 40	54	54	58	64	73	82	89	94	93	89	82	74	70
400	δ	Lupi	S 40 39	04	06	10	15	21	26	29	31	30	27	23	20	19

* No., mag., dist. and p.a. of companion star:　**352**, 4·0, 14″, 152°　　**379**, 1·7, 12″, 225°

No.	Mag.	RA h m	Jan. s	Feb. s	Mar. s	Apr. s	May s	June s	July s	Aug. s	Sept. s	Oct. s	Nov. s	Dec. s	Jan. s
401	3·6	15 21	53·6	54·8	55·9	56·7	57·4	57·8	57·8	57·4	56·8	56·3	56·2	56·5	57·4
402	3·1	15 20	40·7	42·8	45·1	46·9	47·9	47·8	46·7	44·9	42·8	41·0	39·9	39·8	41·0
403	3·7	15 22	46·4	47·7	48·9	49·9	50·6	51·0	51·0	50·5	49·9	49·3	49·1	49·4	50·4
404	4·5	15 24	32·0	33·1	34·1	35·0	35·5	35·7	35·6	35·1	34·4	33·8	33·6	33·7	34·4
405	3·5	15 24	56·2	57·6	59·1	60·3	61·0	61·1	60·6	59·6	58·5	57·4	56·8	56·9	57·8
406	3·7	15 27	52·8	53·8	54·7	55·5	56·1	56·3	56·2	55·8	55·3	54·7	54·5	54·7	55·4
407	4·2	15 32	58·7	59·7	60·7	61·5	62·1	62·3	62·2	61·8	61·2	60·7	60·4	60·6	61·2
408	2·9	15 35	13·8	15·1	16·2	17·2	18·0	18·4	18·4	18·0	17·4	16·8	16·6	16·9	17·8
409	3·8	15 34	51·9	52·9	53·8	54·5	55·1	55·4	55·4	55·1	54·6	54·2	54·0	54·2	54·9
410	2·3	15 34	44·4	45·4	46·4	47·2	47·7	48·0	47·9	47·5	47·0	46·5	46·2	46·4	47·1
411	4·0	15 35	36·2	37·2	38·1	38·9	39·5	39·8	39·9	39·6	39·2	38·8	38·6	38·8	39·6
412	4·1	15 36	50·0	52·1	54·1	55·8	57·1	57·8	57·6	56·6	55·3	54·1	53·6	54·0	55·4
413	3·8	15 37	06·4	07·5	08·5	09·4	10·0	10·4	10·5	10·2	09·7	09·2	09·0	09·3	10·1
414	3·8	15 38	44·4	45·5	46·5	47·4	48·1	48·5	48·5	48·2	47·7	47·2	47·0	47·3	48·1
415	3·9	15 42	47·6	48·6	49·6	50·4	51·0	51·3	51·2	50·8	50·3	49·8	49·5	49·6	50·3
416	2·7	15 44	20·0	20·9	21·8	22·6	23·2	23·5	23·5	23·3	22·8	22·4	22·2	22·4	23·0
417	4·3	15 43	55·3	58·1	61·3	64·1	65·7	65·6	64·1	61·4	58·2	55·1	53·1	52·5	53·7
418	3·7	15 46	14·8	15·8	16·7	17·4	18·0	18·4	18·4	18·1	17·6	17·1	16·9	17·1	17·7
419	4·3	15 48	47·8	48·7	49·6	50·4	51·0	51·4	51·4	51·1	50·6	50·1	49·8	50·0	50·6
420	3·6	15 49	41·4	42·3	43·2	44·0	44·6	45·0	45·1	44·8	44·4	44·0	43·8	44·0	44·6
421	4·1	15 51	02·7	03·8	04·8	05·8	06·5	07·0	07·1	06·8	06·2	05·7	05·5	05·7	06·5
422	3·7	15 50	52·9	53·8	54·7	55·5	56·1	56·5	56·5	56·3	55·8	55·4	55·2	55·4	56·0
423	3·0	15 55	15·1	16·9	18·7	20·4	21·7	22·4	22·3	21·6	20·5	19·4	18·8	19·1	20·3
424	3·9	15 56	30·7	31·6	32·5	33·3	34·0	34·3	34·3	34·1	33·6	33·1	32·8	33·0	33·6
425	4·2	15 57	38·2	39·1	40·1	40·9	41·6	41·9	41·9	41·5	41·0	40·4	40·1	40·2	40·8
426	3·0	15 58	56·0	57·0	58·0	58·9	59·6	60·1	60·2	59·9	59·5	59·0	58·7	58·9	59·6
427	3·6	16 00	12·6	13·7	14·8	15·8	16·7	17·2	17·3	17·0	16·4	15·8	15·5	15·8	16·6
428	2·5	16 00	24·8	25·8	26·7	27·6	28·3	28·8	28·9	28·6	28·2	27·7	27·5	27·7	28·3
429	4·1	16 01	53·1	54·4	55·8	57·1	58·0	58·3	57·9	57·1	55·9	54·7	54·0	53·8	54·4
430	2·9	16 05	30·9	31·9	32·8	33·7	34·4	34·8	35·0	34·7	34·3	33·8	33·6	33·8	34·4
431	4·3	16 06	40·7	41·9	43·0	44·0	44·8	45·3	45·5	45·2	44·6	44·1	43·8	44·0	44·7
432	4·3	16 08	47·8	48·8	49·9	50·9	51·7	52·0	51·9	51·4	50·6	49·8	49·3	49·2	49·8
433	3·0	16 14	24·9	25·8	26·6	27·5	28·1	28·6	28·7	28·5	28·1	27·6	27·4	27·5	28·1
434	4·0	16 15	32·9	34·7	36·5	38·3	39·7	40·5	40·6	40·0	38·9	37·7	37·0	37·2	38·3
435	3·3	16 18	23·4	24·3	25·2	26·0	26·7	27·2	27·3	27·1	26·7	26·3	26·0	26·1	26·7
436	4·1	16 19	56·2	57·5	58·8	60·1	61·1	61·8	61·9	61·6	60·9	60·1	59·7	59·8	60·7
437	3·9	16 19	45·8	46·8	47·9	49·0	49·8	50·1	50·0	49·5	48·7	47·9	47·3	47·2	47·7
438	3·1	16 21	16·1	17·1	18·1	19·0	19·8	20·3	20·5	20·3	19·8	19·3	19·0	19·2	19·8
439	3·8	16 21	58·4	59·3	60·2	61·0	61·7	62·1	62·2	61·9	61·4	60·9	60·6	60·6	61·1
440	2·9	16 23	58·4	59·7	61·2	62·6	63·6	64·0	63·7	62·9	61·5	60·2	59·2	58·9	59·4
441 d	1·2	16 29	29·3	30·3	31·3	32·2	33·0	33·6	33·8	33·6	33·1	32·6	32·3	32·4	33·0
442	2·8	16 30	16·2	17·1	18·0	18·8	19·5	20·0	20·1	19·8	19·3	18·8	18·4	18·4	18·9
443	3·9	16 33	36·4	40·0	44·1	48·0	51·2	53·1	53·2	51·6	48·8	45·9	43·9	43·6	45·6
444	3·8	16 30	58·7	59·5	60·4	61·2	62·0	62·4	62·6	62·4	62·0	61·5	61·2	61·3	61·8
445	4·3	16 31	28·0	29·1	30·2	31·2	32·1	32·7	32·9	32·7	32·2	31·6	31·2	31·3	32·0
446	4·2	16 34	07·8	08·7	09·8	10·8	11·6	12·0	12·0	11·6	10·9	10·1	09·5	09·4	09·8
447	2·9	16 35	57·8	58·8	59·8	60·8	61·6	62·2	62·4	62·2	61·8	61·2	60·9	61·0	61·6
448	2·7	16 37	13·8	14·7	15·6	16·5	17·2	17·7	17·9	17·8	17·4	16·9	16·6	16·7	17·2
449	3·0	16 41	19·5	20·4	21·3	22·2	23·0	23·5	23·5	23·2	22·6	22·0	21·5	21·4	21·9
450	3·6	16 42	55·6	56·5	57·5	58·5	59·3	59·8	59·8	59·4	58·8	58·0	57·5	57·4	57·8

The figures given refer to the beginning of the month, and should be interpolated to the actual date by means of the table on page 73.

No.		Name	Dec	Jan.	F.	M.	Apr.	M.	J.	July	A.	S.	Oct.	N.	D.	Jan.
			° ′	″	″	″	″	″	″	″	″	″	″	″	″	″
401	φ¹	Lupi	S 36 15	54	57	62	66	71	75	78	79	78	76	72	70	70
402	γ	Ursae Minoris	N 71 49	26	19	18	23	31	41	48	51	50	43	33	21	11
403	ε	Lupi	S 44 41	34	36	40	45	51	57	61	63	62	59	54	51	49
404	μ	Bootis	N 37 21	67	59	56	59	65	73	79	83	83	79	71	62	52
405	ι	Draconis	N 58 57	23	15	14	18	26	35	42	46	46	40	30	19	08
406	β	Coronae Bor.	N 29 05	52	44	41	43	48	55	61	64	65	62	55	46	37
407	θ	Coronae Bor.	N 31 20	64	56	53	55	60	67	74	77	78	75	68	59	49
408	γ	Lupi	S 41 10	12	14	17	22	27	32	36	38	37	35	31	27	26
409*	δ	Serpentis	N 10 31	54	47	43	43	46	50	54	57	58	56	52	46	39
410	α	Coronae Bor.	N 26 42	25	17	14	16	21	27	33	37	37	34	28	20	11
411	γ	Librae	S 14 47	40	44	48	51	52	52	52	51	50	49	49	51	54
412	ε	Trianguli Aust.	S 66 19	08	08	11	17	25	33	41	45	45	42	35	28	23
413	υ	Librae	S 28 08	20	23	27	31	34	37	38	39	38	36	34	33	34
414	τ	Librae	S 29 46	54	57	61	65	68	71	73	74	73	71	69	68	69
415	γ	Coronae Bor.	N 26 17	18	10	07	08	13	19	25	29	30	27	21	13	04
416	α	Serpentis	N 6 24	71	64	61	60	62	66	69	72	73	72	69	64	57
417	ζ	Ursae Minoris	N 77 46	67	60	58	62	71	80	88	92	91	85	76	64	53
418	β	Serpentis	N 15 24	55	48	44	44	48	53	58	61	62	60	56	49	41
419	κ	Serpentis	N 18 07	66	58	55	55	59	64	70	73	74	72	67	60	52
420	μ	Serpentis	S 3 26	07	13	17	18	17	16	13	12	10	11	12	16	21
421	χ	Lupi	S 33 37	49	52	55	59	63	66	69	70	70	68	65	63	63
422	ε	Serpentis	N 4 28	20	13	10	09	11	14	17	20	21	20	17	13	06
423	β	Trianguli Aust.	S 63 25	57	56	58	64	71	79	86	91	92	89	83	76	71
424	γ	Serpentis	N 15 39	17	10	06	06	10	15	19	23	24	22	18	11	03
425	ε	Coronae Bor.	N 26 52	16	08	04	05	10	17	23	28	29	26	21	12	03
426	π	Scorpii	S 26 07	03	06	10	13	15	17	19	19	19	17	16	15	16
427	η	Lupi	S 38 23	59	60	63	67	71	75	79	81	81	79	76	73	72
428	δ	Scorpii	S 22 37	31	34	38	41	43	44	45	45	44	43	42	42	44
429	θ	Draconis	N 58 33	27	19	16	19	26	36	44	49	50	46	37	26	15
430	β	Scorpii	S 19 48	33	36	40	42	44	44	45	45	44	43	43	43	45
431	θ	Lupi	S 36 48	18	19	22	26	30	33	37	39	39	37	34	31	30
432	φ	Herculis	N 44 55	41	32	28	30	37	45	53	58	60	57	49	39	28
433	δ	Ophiuchi	S 3 41	55	61	64	66	65	63	60	58	57	57	59	63	67
434	δ	Trianguli Aust.	S 63 41	13	11	12	17	23	31	39	44	45	43	37	30	25
435	ε	Ophiuchi	S 4 41	48	53	56	58	57	55	53	51	50	50	51	55	59
436	γ²	Normae	S 50 09	26	26	28	31	36	42	48	51	52	50	46	41	38
437	τ	Herculis	N 46 18	25	16	12	14	20	29	37	43	44	42	34	24	14
438	σ	Scorpii	S 25 35	44	47	50	52	55	56	58	58	58	57	56	55	56
439	γ	Herculis	N 19 08	52	45	41	41	45	51	56	61	62	61	56	49	41
440	η	Draconis	N 61 30	27	18	14	16	24	33	42	48	50	46	39	28	16
441		*Antares* (α Sco)	S 26 26	05	07	09	12	14	16	17	18	18	17	16	15	15
442	β	Herculis	N 21 28	65	57	53	53	57	63	69	74	76	74	70	62	54
443	γ	Apodis	S 78 53	52	48	48	51	59	68	77	84	87	85	79	71	63
444	λ	Ophiuchi	N 1 58	48	42	38	37	39	42	46	48	50	49	47	43	37
445	N	Scorpii	S 34 42	23	24	27	29	32	36	38	40	41	40	37	35	34
446	σ	Herculis	N 42 25	53	44	40	41	47	55	63	69	72	70	63	54	43
447	τ	Scorpii	S 28 13	06	08	10	13	15	17	19	20	20	19	17	16	16
448	ζ	Ophiuchi	S 10 34	13	17	20	22	22	20	19	18	17	17	18	20	23
449	ζ	Herculis	N 31 35	54	45	41	41	46	54	61	67	69	68	62	54	44
450	η	Herculis	N 38 54	62	53	48	49	55	63	71	77	79	78	72	63	52

* **409:** position refers to midpoint of double star; mags., 4·2 and 5·2, separation, 4″

No.	Mag.	RA	Jan.	Feb.	Mar.	Apr.	May	June	July	Aug.	Sept.	Oct.	Nov.	Dec.	Jan.
		h m	s	s	s	s	s	s	s	s	s	s	s	s	s
451	1·9	16 48	47·1	49·1	51·3	53·5	55·5	56·7	57·1	56·4	55·1	53·6	52·5	52·4	53·4
452	3·7	16 49	53·4	54·9	56·5	58·1	59·5	60·5	60·8	60·4	59·5	58·5	57·8	57·8	58·6
453	4·4	16 45	39·3	42·4	46·9	51·5	54·7	55·6	54·1	50·5	45·4	40·1	35·7	33·3	33·6
454	2·4	16 50	14·8	15·8	16·8	17·9	18·8	19·4	19·7	19·6	19·1	18·5	18·1	18·1	18·7
455	3·1	16 51	57·4	58·5	59·6	60·7	61·6	62·3	62·6	62·5	61·9	61·3	60·9	60·9	61·6
456	3·6	16 52	25·3	26·4	27·5	28·5	29·5	30·2	30·5	30·3	29·8	29·2	28·8	28·8	29·4
457	3·7	16 54	40·4	41·4	42·6	43·7	44·8	45·5	45·8	45·7	45·1	44·4	44·0	44·0	44·6
458	4·3	16 54	03·9	04·7	05·6	06·4	07·2	07·7	07·9	07·8	07·3	06·8	06·4	06·4	06·9
459	3·4	16 57	43·6	44·3	45·2	46·0	46·8	47·3	47·5	47·4	47·0	46·4	46·1	46·0	46·5
460	3·1	16 58	43·2	44·6	46·1	47·6	48·9	49·8	50·2	49·9	49·1	48·2	47·6	47·5	48·3
461	4·1	16 59	40·9	42·2	43·6	45·0	46·3	47·2	47·5	47·3	46·6	45·7	45·1	45·1	45·8
462	3·9	17 00	19·7	20·5	21·5	22·4	23·2	23·7	23·8	23·6	23·0	22·4	21·9	21·7	22·1
463	3·2	17 08	44·4	45·5	47·1	48·8	50·2	50·8	50·7	49·8	48·3	46·6	45·2	44·5	44·6
464	2·6	17 10	27·1	27·9	28·8	29·7	30·5	31·2	31·5	31·4	31·0	30·5	30·1	30·1	30·6
465	3·4	17 12	14·6	15·7	16·8	18·0	19·1	19·9	20·3	20·2	19·6	18·9	18·4	18·4	19·0
466	3·5	17 14	42·1	42·8	43·7	44·5	45·3	45·9	46·1	46·0	45·6	45·0	44·6	44·5	44·9
467	3·2	17 15	04·6	05·3	06·2	07·1	07·9	08·5	08·7	08·5	08·0	07·4	06·9	06·8	07·1
468	3·4	17 15	04·7	05·5	06·4	07·4	08·3	08·8	09·0	08·7	08·1	07·4	06·8	06·6	06·9
469	3·4	17 22	05·3	06·1	07·1	08·0	08·9	09·7	10·0	10·0	09·6	09·0	08·6	08·6	09·0
470	2·8	17 25	24·0	25·2	26·7	28·2	29·6	30·6	31·1	31·0	30·3	29·4	28·6	28·4	29·1
471	3·5	17 25	29·7	30·9	32·4	33·9	35·3	36·4	36·9	36·8	36·0	35·1	34·4	34·2	34·8
472	4·3	17 26	26·9	27·7	28·7	29·6	30·5	31·2	31·6	31·6	31·2	30·7	30·2	30·2	30·6
473	4·4	17 26	34·5	35·2	36·1	36·9	37·7	38·4	38·7	38·6	38·2	37·7	37·3	37·2	37·6
474	4·4	17 27	26·1	27·0	28·0	29·0	30·0	30·7	31·1	31·1	30·7	30·1	29·7	29·6	30·1
475	3·8	17 31	12·2	13·6	15·2	16·9	18·5	19·8	20·3	20·1	19·3	18·2	17·4	17·1	17·7
476	2·8	17 30	51·0	51·9	53·0	54·0	55·1	55·9	56·3	56·3	55·9	55·2	54·8	54·7	55·1
477	4·5	17 30	46·9	47·6	48·4	49·3	50·2	50·8	51·0	50·9	50·4	49·8	49·3	49·1	49·4
478	3·0	17 30	26·0	26·8	28·0	29·2	30·2	30·9	31·0	30·6	29·7	28·7	27·8	27·3	27·5
479	3·0	17 31	56·2	57·3	58·5	59·9	61·1	62·1	62·6	62·5	61·9	61·2	60·5	60·4	60·9
480	1·7	17 33	41·7	42·6	43·6	44·7	45·8	46·6	47·0	47·0	46·6	45·9	45·5	45·3	45·8
481	2·1	17 34	59·4	60·0	60·9	61·7	62·5	63·2	63·5	63·4	63·0	62·5	62·0	61·9	62·2
482	2·0	17 37	24·5	25·5	26·6	27·8	28·9	29·8	30·3	30·3	29·8	29·1	28·6	28·4	28·9
483	3·6	17 37	39·5	40·3	41·2	42·1	42·9	43·6	44·0	44·0	43·7	43·2	42·8	42·7	43·0
484	3·8	17 39	28·8	29·5	30·5	31·6	32·6	33·3	33·5	33·2	32·5	31·6	30·8	30·4	30·5
485	2·5	17 42	34·5	35·4	36·4	37·5	38·6	39·5	40·0	40·0	39·5	38·9	38·4	38·3	38·7
486	2·9	17 43	31·9	32·6	33·4	34·3	35·1	35·8	36·1	36·1	35·7	35·2	34·8	34·7	35·0
487	3·6	17 45	50·6	52·1	53·9	55·8	57·7	59·1	59·9	59·7	58·8	57·5	56·5	56·0	56·6
488	3·5	17 46	29·9	30·6	31·4	32·3	33·2	33·8	34·1	34·0	33·5	32·9	32·3	32·1	32·3
489	3·1	17 47	40·3	41·2	42·3	43·4	44·5	45·4	45·9	45·9	45·5	44·9	44·3	44·2	44·6
490	3·7	17 47	57·2	57·9	58·7	59·6	60·4	61·1	61·4	61·4	61·1	60·6	60·1	60·0	60·3
491	3·2	17 49	56·6	57·5	58·5	59·6	60·7	61·5	62·0	62·1	61·6	61·0	60·5	60·4	60·8
492	3·9	17 53	31·0	31·7	32·8	34·2	35·4	36·2	36·4	36·0	35·0	33·8	32·7	32·0	32·0
493	4·0	17 56	16·9	17·5	18·4	19·4	20·3	21·0	21·3	21·2	20·6	19·9	19·2	18·9	19·0
494	2·4	17 56	36·5	37·2	38·2	39·4	40·5	41·3	41·5	41·2	40·4	39·4	38·4	37·9	37·9
495	3·8	17 57	48·2	48·8	49·7	50·6	51·5	52·2	52·5	52·4	51·9	51·3	50·7	50·4	50·6
496	3·5	17 59	05·7	06·4	07·2	08·1	09·0	09·7	10·1	10·2	09·9	09·4	08·9	08·8	09·1
497	3·9	18 00	42·4	43·0	43·8	44·6	45·5	46·2	46·6	46·6	46·3	45·8	45·3	45·2	45·4
498	3·1	18 05	53·3	54·1	55·0	56·0	57·0	57·9	58·4	58·5	58·2	57·6	57·1	56·9	57·3
499	3·9	18 06	43·5	44·5	45·7	47·0	48·4	49·5	50·1	50·2	49·7	48·9	48·2	47·9	48·3
500	3·7	18 07	24·3	24·9	25·7	26·6	27·4	28·1	28·5	28·5	28·2	27·7	27·2	27·0	27·2

The figures given refer to the beginning of the month, and should be interpolated to the actual date by means of the table on page 73.

No.		Name	Dec	Jan.	F.	M.	Apr.	M.	J.	July	A.	S.	Oct.	N.	D.	Jan.
			° ′	″	″	″	″	″	″	″	″	″	″	″	″	″
451	α	Trianguli Aust.	S 69 01	42	38	38	41	47	54	63	69	72	71	66	58	51
452	η	Arae	S 59 02	32	29	30	32	37	44	51	56	58	57	53	47	41
453	ε	Ursae Minoris	N 82 01	54	45	41	43	50	59	68	74	76	73	66	56	45
454	ε	Scorpii	S 34 17	42	43	45	47	49	52	55	57	58	57	55	53	52
455	μ¹	Scorpii	S 38 02	56	56	58	60	63	66	69	72	73	72	70	67	65
456	μ²	Scorpii	S 38 01	09	09	10	12	15	19	22	25	26	25	23	20	18
457	ζ²	Scorpii	S 42 21	46	45	47	49	52	56	60	64	65	64	61	58	55
458	ι	Ophiuchi	N 10 09	43	36	32	31	34	39	44	47	49	49	46	41	34
459	κ	Ophiuchi	N 9 22	18	12	08	07	10	14	19	23	25	25	22	16	10
460	ζ	Arae	S 55 59	27	25	25	27	31	37	44	49	51	50	46	41	36
461	ε¹	Arae	S 53 09	41	38	38	41	45	50	56	61	63	63	59	54	49
462	ε	Herculis	N 30 55	21	12	08	08	13	20	27	33	36	35	30	22	13
463	ζ	Draconis	N 65 42	38	28	23	23	30	39	48	56	59	58	52	42	31
464	η	Ophiuchi	S 15 43	36	39	41	43	43	42	41	41	40	40	40	41	43
465	η	Scorpii	S 43 14	24	23	24	25	28	32	36	40	42	41	39	35	32
466*	α	Herculis	N 14 23	16	09	04	04	07	12	18	23	25	25	22	16	09
467	δ	Herculis	N 24 49	70	62	57	57	61	67	74	80	83	83	79	72	63
468	π	Herculis	N 36 48	21	12	07	07	12	19	28	34	38	37	32	24	14
469	θ	Ophiuchi	S 25 00	02	04	05	06	07	08	09	10	10	10	09	09	09
470	β	Arae	S 55 31	48	45	44	45	49	54	60	65	69	69	65	60	55
471	γ	Arae	S 56 22	41	37	36	37	41	46	52	58	61	61	58	52	47
472	44	Ophiuchi	S 24 10	35	36	38	39	39	40	41	41	42	42	41	41	41
473	σ	Ophiuchi	N 4 08	19	13	09	08	11	15	19	22	24	24	22	18	13
474	45	Ophiuchi	S 29 52	05	05	06	07	08	10	11	13	14	14	13	12	11
475	δ	Arae	S 60 40	62	57	56	57	60	66	73	79	83	83	80	74	68
476	υ	Scorpii	S 37 17	47	46	46	47	49	52	55	57	59	59	58	55	53
477	λ	Herculis	N 26 06	30	22	17	16	21	27	35	41	44	44	40	33	25
478	β	Draconis	N 52 17	55	45	39	39	45	53	63	71	75	75	70	61	50
479	α	Arae	S 49 52	35	32	31	32	35	40	45	49	52	52	50	45	41
480	λ	Scorpii	S 37 06	15	15	15	16	17	20	23	25	27	27	26	23	21
481	α	Ophiuchi	N 12 33	30	23	18	18	21	26	31	36	39	39	36	31	24
482	θ	Scorpii	S 42 59	53	51	51	51	53	57	61	64	67	67	65	62	58
483	ξ	Serpentis	S 15 23	58	61	63	64	63	62	61	60	60	60	61	61	63
484	ι	Herculis	N 45 59	75	65	59	59	64	72	81	89	94	94	89	81	71
485	κ	Scorpii	S 39 01	49	47	47	48	49	52	55	58	60	60	59	56	53
486	β	Ophiuchi	N 4 33	58	53	49	48	51	55	59	63	65	65	63	59	54
487	η	Pavonis	S 64 43	25	19	17	17	20	26	33	40	45	45	42	36	29
488	μ	Herculis	N 27 42	68	59	53	53	57	64	71	78	81	82	78	71	63
489	ι¹	Scorpii	S 40 07	38	36	35	36	37	40	43	46	49	49	48	45	42
490	γ	Ophiuchi	N 2 42	23	17	14	13	15	19	23	27	29	29	27	23	19
491	G	Scorpii	S 37 02	36	35	34	35	36	38	41	43	45	46	45	42	40
492	ξ	Draconis	N 56 51	77	66	60	59	64	72	82	91	96	97	93	84	73
493	θ	Herculis	N 37 14	58	48	43	42	46	54	62	70	75	75	72	64	55
494	γ	Draconis	N 51 28	76	66	59	58	63	71	81	89	95	96	92	83	73
495	ξ	Herculis	N 29 14	49	40	35	34	38	45	53	60	64	65	61	55	46
496	ν	Ophiuchi	S 9 46	26	30	32	32	31	29	27	25	24	24	25	27	29
497	67	Ophiuchi	N 2 55	52	47	43	43	45	49	53	57	59	59	58	54	49
498	γ	Sagittarii	S 30 25	26	26	26	26	26	26	28	30	31	32	31	30	29
499	θ	Arae	S 50 05	28	24	22	21	23	26	31	36	40	41	39	35	30
500	72	Ophiuchi	N 9 33	49	43	39	38	41	46	51	56	59	59	57	53	47

* No., mag., dist. and p.a. of companion star: **466**, 5·4, 5″, 104°

No.	Mag.	RA	Jan.	Feb.	Mar.	Apr.	May	June	July	Aug.	Sept.	Oct.	Nov.	Dec.	Jan.
		h m	s	s	s	s	s	s	s	s	s	s	s	s	s
501	3·8	18 07	34·9	35·5	36·3	37·2	38·1	38·8	39·2	39·1	38·7	38·0	37·4	37·1	37·3
502	4·0	18 13	50·3	51·0	51·8	52·8	53·7	54·5	55·1	55·1	54·9	54·4	53·9	53·7	54·0
503	3·2	18 17	42·7	43·5	44·4	45·5	46·6	47·6	48·1	48·3	47·9	47·3	46·8	46·5	46·8
504	4·3	18 19	53·5	54·1	54·9	55·8	56·8	57·6	57·9	57·9	57·4	56·7	56·0	55·6	55·6
505	2·8	18 21	04·5	05·2	06·1	07·1	08·1	09·0	09·6	09·7	09·4	08·8	08·3	08·1	08·4
506	3·4	18 21	22·4	23·0	23·7	24·6	25·5	26·2	26·7	26·8	26·5	26·0	25·5	25·3	25·6
507	4·2	18 23	19·9	21·0	22·6	24·3	26·1	27·6	28·4	28·5	27·9	26·8	25·8	25·3	25·6
508	3·7	18 20	57·0	57·7	59·5	61·7	63·8	65·2	65·5	64·7	62·8	60·4	58·1	56·4	55·8
509	1·9	18 24	15·3	16·0	16·9	18·0	19·1	20·0	20·6	20·7	20·4	19·9	19·3	19·1	19·4
510	3·9	18 23	44·7	45·2	46·0	46·8	47·7	48·5	48·9	48·9	48·6	48·0	47·4	47·1	47·3
511	3·8	18 27	03·8	04·7	05·8	07·0	08·2	09·3	10·0	10·2	09·8	09·1	08·5	08·1	08·4
512	2·9	18 28	02·9	03·5	04·4	05·3	06·4	07·2	07·8	07·9	07·7	07·1	06·7	06·4	06·7
513	4·1	18 35	16·5	17·0	17·8	18·7	19·6	20·4	20·9	21·0	20·8	20·3	19·9	19·6	19·8
514 d	0·1	18 36	57·9	58·4	59·2	60·2	61·2	62·0	62·5	62·4	61·9	61·2	60·5	60·0	60·0
515	4·1	18 43	09·4	10·8	12·9	15·4	18·0	20·2	21·5	21·7	20·8	19·2	17·5	16·5	16·6
516	3·3	18 45	44·1	44·7	45·5	46·5	47·5	48·4	49·0	49·2	49·0	48·5	48·0	47·7	47·9
517	4·3	18 45	42·6	43·0	43·7	44·6	45·5	46·3	46·8	46·8	46·5	46·0	45·4	45·1	45·2
518	4·5	18 47	14·4	14·9	15·6	16·5	17·4	18·2	18·7	18·9	18·7	18·2	17·8	17·5	17·7
519	4·4	18 47	04·2	04·7	05·4	06·2	07·2	07·9	08·4	08·5	08·2	07·7	07·2	06·8	06·9
520	3·8	18 50	06·8	07·3	08·0	08·9	09·9	10·7	11·2	11·2	10·9	10·2	09·5	09·1	09·1
521	4·4	18 52	19·3	20·3	21·7	23·4	25·3	26·9	28·0	28·2	27·7	26·7	25·6	24·9	25·0
522	2·1	18 55	20·6	21·2	22·0	22·9	23·9	24·9	25·5	25·7	25·5	25·0	24·5	24·2	24·4
523	4·2	18 55	21·2	21·6	22·4	23·4	24·5	25·4	25·9	25·9	25·4	24·6	23·8	23·2	23·1
524	3·6	18 57	48·3	48·8	49·6	50·5	51·5	52·4	53·0	53·2	53·0	52·6	52·1	51·8	52·0
525	3·3	18 58	58·7	59·1	59·8	60·7	61·7	62·6	63·1	63·1	62·8	62·2	61·5	61·0	61·0
526	4·2	18 59	40·5	40·9	41·6	42·4	43·3	44·1	44·7	44·8	44·5	44·1	43·5	43·2	43·2
527	2·7	19 02	41·5	42·0	42·8	43·8	44·9	45·9	46·5	46·8	46·6	46·1	45·6	45·3	45·4
528	3·9	19 04	45·5	46·0	46·8	47·7	48·7	49·6	50·2	50·5	50·3	49·9	49·4	49·1	49·2
529	3·0	19 05	27·8	28·2	28·9	29·7	30·6	31·4	32·0	32·1	31·9	31·4	30·9	30·6	30·6
530	3·5	19 06	18·9	19·3	20·0	20·8	21·8	22·6	23·2	23·4	23·2	22·8	22·3	22·0	22·1
531	3·4	19 07	01·1	01·7	02·4	03·4	04·4	05·4	06·1	06·3	06·1	05·7	05·1	04·8	05·0
532	4·1	19 09	33·4	34·0	34·9	35·9	37·1	38·1	38·9	39·2	39·0	38·5	37·9	37·5	37·6
533	3·0	19 09	50·3	50·8	51·6	52·4	53·4	54·4	55·0	55·3	55·1	54·7	54·2	53·9	54·0
534	3·2	19 12	29·7	29·9	31·1	32·7	34·6	36·0	36·7	36·4	35·3	33·7	31·8	30·3	29·5
535	4·5	19 16	23·8	24·2	24·8	25·8	26·8	27·7	28·3	28·4	28·0	27·4	26·6	26·0	25·9
536	4·0	19 17	06·2	06·5	07·3	08·4	09·7	10·7	11·3	11·3	10·7	09·8	08·7	07·8	07·5
537	3·9	19 21	44·7	45·2	45·9	46·7	47·7	48·6	49·3	49·6	49·5	49·0	48·5	48·2	48·3
538	4·3	19 22	43·6	44·2	45·1	46·2	47·4	48·7	49·5	49·9	49·7	49·1	48·4	48·0	48·0
539	4·1	19 23	58·3	58·9	59·7	60·8	62·0	63·1	63·9	64·3	64·1	63·6	63·0	62·5	62·6
540	3·4	19 25	33·6	34·0	34·6	35·4	36·3	37·2	37·8	38·0	37·9	37·5	37·0	36·7	36·7
541	3·9	19 29	42·7	42·9	43·6	44·7	46·0	47·0	47·7	47·8	47·3	46·4	45·3	44·5	44·1
542	3·2	19 30	45·8	46·1	46·7	47·6	48·5	49·4	50·0	50·2	50·0	49·4	48·8	48·3	48·2
543	4·4	19 41	06·0	06·3	06·9	07·7	08·6	09·5	10·1	10·4	10·2	09·8	09·2	08·8	08·8
544	3·0	19 44	59·8	60·0	60·6	61·5	62·7	63·7	64·4	64·5	64·2	63·5	62·6	61·9	61·6
545	2·8	19 46	19·0	19·3	19·9	20·6	21·5	22·4	23·1	23·4	23·2	22·8	22·3	21·9	21·9
546	3·8	19 47	26·3	26·6	27·2	27·9	28·9	29·8	30·4	30·7	30·5	30·1	29·6	29·1	29·0
547	4·0	19 48	06·1	06·0	07·0	08·7	10·8	12·5	13·5	13·5	12·5	10·7	08·6	06·8	05·6
548 d	0·9	19 50	50·6	50·9	51·4	52·2	53·1	54·0	54·6	54·9	54·9	54·5	54·0	53·6	53·5
549	4·0	19 52	32·2	32·5	33·0	33·8	34·7	35·6	36·3	36·6	36·5	36·2	35·7	35·3	35·3
550	4·2	19 55	20·8	21·3	22·0	23·0	24·2	25·5	26·4	26·9	26·8	26·3	25·7	25·2	25·1

The figures given refer to the beginning of the month, and should be interpolated to the actual date by means of the table on page 73.

No.		Name	Dec	Jan.	F.	M.	Apr.	M.	J.	July	A.	S.	Oct.	N.	D.	Jan.
			° ′	″	″	″	″	″	″	″	″	″	″	″	″	″
501	o	Herculis	N 28 45	44	35	29	28	32	39	47	54	58	59	56	50	41
502	μ	Sagittarii	S 21 03	30	31	32	32	31	31	30	31	31	31	32	32	32
503	η	Sagittarii	S 36 45	40	38	37	37	37	38	41	44	46	47	46	44	42
504	κ	Lyrae	N 36 03	54	44	38	36	40	47	56	64	69	71	68	61	52
505	δ	Sagittarii	S 29 49	39	38	38	38	37	37	39	40	42	43	43	41	40
506	η	Serpentis	S 2 53	55	60	62	63	61	57	54	51	50	49	51	53	57
507	ξ	Pavonis	S 61 29	34	29	25	24	25	29	35	42	47	49	47	42	35
508	χ	Draconis	N 72 43	59	48	41	39	43	51	61	70	77	79	76	68	58
509	ε	Sagittarii	S 34 22	62	61	60	59	59	60	62	64	66	67	67	65	63
510	109	Herculis	N 21 45	72	64	59	58	61	67	75	81	85	87	84	79	71
511	α	Telescopii	S 45 57	63	59	57	56	57	59	63	67	71	72	71	68	64
512	λ	Sagittarii	S 25 25	16	16	16	15	14	14	14	15	16	17	17	17	16
513	α	Scuti	S 8 14	35	39	41	41	39	36	33	31	30	30	31	33	36
514	Vega (α Lyrae)		N 38 46	66	56	50	48	51	59	68	76	82	84	82	76	67
515	ζ	Pavonis	S 71 25	36	29	23	21	22	27	34	42	48	51	50	44	36
516	φ	Sagittarii	S 26 59	22	21	21	20	19	18	18	19	21	22	22	22	21
517	110	Herculis	N 20 32	52	44	39	37	40	46	54	60	65	66	64	59	53
518	β	Scuti	S 4 44	47	51	53	53	51	47	44	41	40	39	40	43	46
519	111	Herculis	N 18 10	59	52	47	46	49	55	62	68	72	74	72	67	61
520	β	Lyrae	N 33 21	53	43	37	35	38	45	53	62	68	70	68	62	54
521	λ	Pavonis	S 62 10	70	63	58	56	56	59	65	72	78	81	80	75	68
522	σ	Sagittarii	S 26 17	42	42	41	40	39	38	38	39	40	41	42	41	41
523	R	Lyrae	N 43 56	54	44	37	34	38	45	54	63	70	73	72	66	57
524	ξ²	Sagittarii	S 21 06	18	18	18	17	15	14	13	13	14	14	15	15	15
525	γ	Lyrae	N 32 41	31	22	15	13	16	23	32	40	46	48	47	41	33
526	ε	Aquilae	N 15 04	14	07	02	01	04	09	16	22	26	27	26	22	16
527	ζ	Sagittarii	S 29 52	43	41	40	39	37	36	37	38	40	42	42	42	40
528	o	Sagittarii	S 21 44	23	23	23	22	20	18	18	18	18	19	20	20	20
529	ζ	Aquilae	N 13 51	57	50	46	45	47	53	59	65	69	71	70	66	60
530	λ	Aquilae	S 4 52	49	53	55	55	52	49	45	42	41	40	41	44	47
531	τ	Sagittarii	S 27 40	07	06	05	04	02	01	01	02	04	05	06	05	04
532	α	Coronae Aust.	S 37 54	09	07	04	02	00	00	02	05	08	10	11	09	06
533	π	Sagittarii	S 21 01	18	18	18	16	14	12	11	11	12	13	14	14	14
534	δ	Draconis	N 67 39	55	44	35	31	34	41	51	61	69	74	73	68	58
535	θ	Lyrae	N 38 07	74	64	57	54	57	64	73	81	88	92	91	86	78
536	κ	Cygni	N 53 21	80	69	61	58	60	67	77	87	95	99	99	93	84
537	ρ	Sagittarii	S 17 50	41	42	42	41	38	36	34	33	33	34	35	36	37
538	β¹	Sagittarii	S 44 27	25	20	17	14	12	12	15	18	23	26	26	24	20
539	α	Sagittarii	S 40 36	50	46	43	40	38	38	40	43	47	50	50	49	45
540	δ	Aquilae	N 3 06	64	60	57	56	59	64	69	73	76	77	76	73	69
541	ι	Cygni	N 51 43	63	53	44	41	43	50	59	69	78	82	82	77	69
542	β	Cygni	N 27 57	49	40	34	32	34	41	49	57	63	66	65	61	54
543	β	Sagittae	N 17 28	48	42	37	35	37	43	50	57	62	64	64	60	54
544	δ	Cygni	N 45 07	69	59	51	47	49	56	65	74	83	88	88	83	75
545	γ	Aquilae	N 10 36	63	57	53	52	54	60	66	71	76	78	77	74	69
546	δ	Sagittae	N 18 32	19	12	07	05	08	13	21	27	33	36	35	32	26
547	ε	Draconis	N 70 16	25	15	05	00	01	07	17	28	37	43	44	40	32
548	Altair (α Aql)		N 8 52	22	17	13	12	15	20	26	31	36	37	37	34	29
549	η	Aquilae	N 1 00	35	31	28	28	31	35	40	45	48	49	48	45	42
550	ι	Sagittarii	S 41 51	56	51	47	43	40	39	41	44	48	51	53	52	48

No.	Mag.	RA	Jan.	Feb.	Mar.	Apr.	May	June	July	Aug.	Sept.	Oct.	Nov.	Dec.	Jan.
		h m	s	s	s	s	s	s	s	s	s	s	s	s	s
551	3·9	19 55	22·4	22·7	23·2	24·0	24·9	25·8	26·5	26·8	26·7	26·3	25·8	25·5	25·4
552	4·0	19 56	20·5	20·7	21·3	22·1	23·1	24·1	24·8	25·0	24·8	24·3	23·6	23·0	22·8
553	3·7	19 58	48·5	48·7	49·3	50·0	50·9	51·9	52·5	52·8	52·7	52·3	51·8	51·3	51·2
554	4·1	20 00	42·5	43·2	44·9	47·2	50·0	52·8	54·8	55·7	55·4	54·0	52·1	50·5	49·9
555	4·4	19 59	49·1	49·5	50·2	51·1	52·2	53·3	54·2	54·7	54·6	54·2	53·6	53·2	53·1
556	3·6	20 08	50·2	50·8	52·0	53·7	55·8	57·9	59·5	60·3	60·1	59·2	57·9	56·8	56·4
557	4·4	20 08	43·8	43·2	44·3	46·8	49·9	52·6	54·2	54·3	52·8	50·0	46·6	43·4	41·2
558	3·4	20 11	22·2	22·4	22·9	23·7	24·6	25·5	26·2	26·6	26·6	26·3	25·8	25·4	25·3
559	4·3	20 13	23·9	23·9	24·4	25·5	26·9	28·1	29·0	29·3	28·9	28·0	26·8	25·8	25·2
560	3·9	20 13	39·2	39·3	39·8	40·7	41·8	42·9	43·7	44·0	43·8	43·1	42·2	41·5	41·0
561	3·8	20 18	07·5	07·8	08·3	09·1	10·0	11·0	11·7	12·2	12·2	11·9	11·4	11·0	11·0
562	3·2	20 21	05·0	05·3	05·8	06·5	07·5	08·4	09·2	09·7	09·7	09·4	08·9	08·6	08·5
563	2·3	20 22	15·7	15·7	16·2	17·0	18·1	19·1	19·9	20·2	20·1	19·5	18·8	18·1	17·8
564	2·1	20 25	44·5	44·8	45·7	46·9	48·4	50·0	51·4	52·1	52·1	51·5	50·6	49·8	49·5
565	4·1	20 29	26·4	26·5	26·9	27·7	28·6	29·6	30·4	30·8	30·7	30·3	29·7	29·1	28·9
566	4·3	20 29	33·9	33·6	34·2	35·3	36·9	38·4	39·5	39·8	39·4	38·3	36·9	35·6	34·7
567	4·0	20 33	16·3	16·5	16·9	17·6	18·5	19·4	20·2	20·6	20·6	20·3	19·8	19·4	19·2
568	3·2	20 37	39·3	39·6	40·2	41·2	42·5	43·8	44·9	45·6	45·7	45·3	44·6	44·0	43·7
569	3·7	20 37	36·4	36·5	37·0	37·6	38·5	39·5	40·2	40·6	40·7	40·4	39·9	39·4	39·2
570	3·9	20 39	41·7	41·8	42·2	42·9	43·8	44·7	45·5	45·9	46·0	45·7	45·2	44·7	44·5
571 d	1·3	20 41	27·7	27·6	28·0	28·8	29·9	31·0	31·9	32·3	32·2	31·6	30·8	30·1	29·6
572	3·6	20 45	03·5	03·7	04·7	06·2	08·2	10·3	12·0	13·1	13·1	12·4	11·1	09·9	09·3
573	4·3	20 46	10·4	10·6	11·0	11·8	12·8	13·8	14·7	15·3	15·4	15·1	14·6	14·2	14·0
574	3·6	20 45	16·9	16·6	17·1	18·1	19·7	21·2	22·3	22·7	22·4	21·5	20·1	18·8	17·9
575	2·6	20 46	15·3	15·4	15·7	16·4	17·4	18·5	19·3	19·7	19·7	19·3	18·7	18·1	17·7
576	3·8	20 47	44·8	45·0	45·4	46·1	46·9	47·9	48·8	49·3	49·4	49·1	48·7	48·3	48·1
577	4·2	20 51	54·0	54·2	54·6	55·3	56·3	57·4	58·3	58·9	59·0	58·8	58·3	57·9	57·7
578	3·7	20 54	54·2	54·4	55·1	56·2	57·8	59·5	60·9	61·8	61·9	61·4	60·5	59·6	59·2
579	4·0	20 57	12·7	12·6	12·9	13·6	14·7	15·8	16·6	17·1	17·1	16·7	16·0	15·3	14·8
580	3·9	21 04	58·0	57·9	58·2	58·9	59·9	61·1	62·0	62·5	62·5	62·0	61·3	60·6	60·1
581	4·2	21 06	01·3	01·4	01·8	02·5	03·4	04·4	05·3	05·9	06·0	05·8	05·4	05·0	04·8
582	3·4	21 12	59·2	59·2	59·5	60·1	61·0	62·0	62·9	63·4	63·5	63·2	62·6	62·1	61·7
583	3·8	21 14	50·1	50·0	50·3	50·9	51·9	53·0	53·9	54·4	54·5	54·2	53·5	52·9	52·5
584	4·1	21 15	53·4	53·4	53·7	54·3	55·2	56·1	57·0	57·5	57·7	57·5	ʻ57·1	56·6	56·4
585	4·3	21 17	27·5	27·4	27·7	28·3	29·3	30·4	31·3	31·9	31·9	31·6	30·9	30·3	29·8
586	4·4	21 17	57·9	57·8	58·1	58·7	59·7	60·7	61·6	62·1	62·2	61·9	61·3	60·7	60·3
587	2·6	21 18	34·9	34·4	34·6	35·6	37·0	38·6	39·9	40·5	40·4	39·7	38·4	37·1	36·0
588	4·3	21 22	19·3	19·4	19·7	20·3	21·2	22·2	23·1	23·7	23·9	23·8	23·4	23·0	22·7
589	4·3	21 22	08·7	08·7	09·0	09·6	10·4	11·4	12·3	12·8	12·9	12·7	12·3	11·8	11·5
590	4·3	21 26	32·3	32·3	32·9	34·1	35·9	37·9	39·8	41·0	41·3	40·8	39·7	38·5	37·7
591	3·9	21 26	44·6	44·7	45·0	45·7	46·6	47·6	48·5	49·2	49·4	49·3	48·9	48·4	48·2
592	3·3	21 28	37·8	36·9	37·0	38·2	40·1	42·2	43·9	44·7	44·6	43·5	41·8	39·8	38·2
593	3·1	21 31	37·8	37·8	38·1	38·7	39·5	40·5	41·4	42·0	42·2	42·0	41·7	41·3	41·0
594	4·2	21 34	01·2	01·0	01·2	01·8	02·8	04·0	05·0	05·6	05·7	05·4	04·7	03·9	03·4
595	3·8	21 40	10·0	10·0	10·3	10·9	11·7	12·7	13·7	14·3	14·6	14·5	14·1	13·7	13·4
596	3·7	21 41	35·0	34·4	35·2	37·2	40·3	43·9	47·2	49·5	50·2	49·2	47·0	44·4	42·4
597	4·4	21 43	31·7	31·2	31·3	32·0	33·3	34·8	36·0	36·8	36·9	36·4	35·4	34·3	33·3
598	2·5	21 44	15·1	15·1	15·3	15·8	16·6	17·6	18·5	19·1	19·3	19·2	18·8	18·4	18·1
599	4·3	21 45	01·6	01·7	02·0	02·6	03·5	04·6	05·7	06·4	06·8	06·6	06·2	05·7	05·4
600	4·5	21 45	27·9	27·3	27·4	28·1	29·5	31·0	32·4	33·2	33·3	32·7	31·6	30·4	29·4

The figures given refer to the beginning of the month, and should be interpolated to the actual date by means of the table on page 73.

No.		Name	Dec	Jan.	F.	M.	Apr.	M.	J.	July	A.	S.	Oct.	N.	D.	Jan.
			° ′	″	″	″	″	″	″	″	″	″	″	″	″	″
551	β	Aquilae	N 6 24	39	34	30	30	32	37	43	48	52	53	53	50	46
552	η	Cygni	N 35 05	20	11	03	00	02	08	17	26	33	38	38	34	27
553	γ	Sagittae	N 19 29	50	43	37	35	38	43	51	58	63	66	66	63	57
554	ε	Pavonis	S 72 54	29	20	12	06	03	05	10	18	26	31	33	29	22
555	θ¹	Sagittarii	S 35 16	24	21	17	14	11	09	09	11	14	17	19	18	16
556	δ	Pavonis	S 66 10	48	40	32	27	24	24	28	35	43	48	50	47	41
557	κ	Cephei	N 77 42	67	56	47	41	41	46	55	66	76	83	85	82	75
558	θ	Aquilae	S 0 48	61	64	66	66	63	59	54	50	47	46	47	49	52
559	33	Cygni	N 56 34	29	18	09	04	05	11	20	31	40	47	48	45	38
560	o²	Cygni	N 46 44	53	43	34	30	31	37	46	56	65	71	72	69	62
561	α²	Capricorni	S 12 32	26	27	27	26	22	18	15	13	12	13	14	15	17
562	β	Capricorni	S 14 46	37	38	38	36	33	29	26	24	24	24	26	27	28
563	γ	Cygni	N 40 15	49	39	32	27	29	34	43	52	61	67	68	65	58
564	α	Pavonis	S 56 43	56	49	42	36	32	31	33	39	45	50	53	51	46
565	41	Cygni	N 30 22	31	23	16	13	14	20	28	37	44	49	50	47	41
566	θ	Cephei	N 62 59	67	57	47	42	41	46	55	66	76	84	87	84	77
567	ε	Delphini	N 11 18	33	28	24	23	25	30	37	43	48	50	50	48	44
568	α	Indi	S 47 16	77	71	66	60	56	53	54	58	63	68	70	70	66
569	β	Delphini	N 14 35	65	59	55	53	55	61	67	74	80	83	83	80	76
570	α	Delphini	N 15 54	66	60	56	54	56	61	68	75	81	84	84	82	77
571	Deneb (α Cygni)		N 45 16	78	68	60	55	55	61	69	79	89	95	98	95	89
572	β	Pavonis	S 66 11	60	52	43	36	31	30	33	39	47	53	56	55	49
573	ψ	Capricorni	S 25 15	59	58	56	52	48	44	42	42	43	46	48	49	48
574	η	Cephei	N 61 50	53	43	33	27	26	31	40	51	61	69	73	71	65
575	ε	Cygni	N 33 58	41	33	26	22	23	28	36	45	53	59	61	58	53
576	ε	Aquarii	S 9 29	26	27	28	26	23	18	14	11	10	10	11	13	14
577	ω	Capricorni	S 26 54	53	51	48	44	40	36	34	34	36	38	41	42	41
578	β	Indi	S 58 26	63	56	48	41	36	34	35	40	47	53	56	56	51
579	ν	Cygni	N 41 10	32	23	15	11	11	16	24	34	43	49	52	50	44
580	ξ	Cygni	N 43 55	72	63	55	50	50	54	63	73	82	89	92	91	85
581	θ	Capricorni	S 17 13	39	39	38	35	31	27	23	21	21	22	24	26	26
582	ζ	Cygni	N 30 13	67	60	53	50	50	55	63	71	79	85	87	86	81
583	τ	Cygni	N 38 02	77	69	61	57	57	61	69	79	88	95	97	96	91
584	α	Equulei	N 5 15	17	13	11	10	13	18	24	29	34	36	36	34	31
585	σ	Cygni	N 39 23	74	66	58	53	53	57	65	75	84	91	94	93	87
586	υ	Cygni	N 34 54	21	13	06	01	01	06	14	23	32	38	41	39	34
587	α	Cephei	N 62 35	45	36	26	19	17	21	29	39	50	59	64	64	59
588	ι	Capricorni	S 16 49	44	44	43	40	36	31	27	25	24	26	28	29	30
589	I	Pegasi	N 19 48	46	40	35	32	33	38	45	53	60	64	65	64	60
590	γ	Pavonis	S 65 21	44	36	27	19	12	09	10	15	23	30	34	34	29
591	ζ	Capricorni	S 22 24	21	20	18	14	09	05	01	00	01	03	05	07	07
592	β	Cephei	N 70 33	78	69	59	52	49	52	59	70	81	90	96	97	92
593	β	Aquarii	S 5 33	52	54	55	54	50	45	40	36	34	33	34	36	38
594	ρ	Cygni	N 45 35	67	59	50	44	43	47	55	65	75	82	87	86	81
595	γ	Capricorni	S 16 39	23	23	21	18	14	09	04	02	02	03	05	07	08
596	ν	Octantis	S 77 22	72	63	53	43	36	33	35	41	50	58	63	63	57
597	μ	Cephei	N 58 47	28	19	10	03	00	03	11	21	32	41	47	47	43
598	ε	Pegasi	N 9 52	59	55	51	50	52	57	64	70	75	78	79	77	74
599	ι	Piscis Aust.	S 33 00	75	72	68	62	56	51	49	49	51	55	58	60	60
600	ν	Cephei	N 61 07	55	47	37	30	27	30	38	48	59	68	74	75	71

No.	Mag.	RA h m	Jan. s	Feb. s	Mar. s	Apr. s	May s	June s	July s	Aug. s	Sept. s	Oct. s	Nov. s	Dec. s	Jan. s
601	3·0	21 47	07·0	07·0	07·3	07·8	08·6	09·6	10·6	11·3	11·5	11·5	11·1	10·7	10·4
602	4·3	21 46	49·9	49·5	49·7	50·3	51·4	52·6	53·7	54·4	54·5	54·2	53·5	52·6	52·0
603	3·2	21 54	00·6	00·6	00·9	01·5	02·5	03·6	04·7	05·6	05·9	05·8	05·4	04·8	04·5
604	3·2	22 05	51·3	51·3	51·4	51·9	52·7	53·6	54·5	55·2	55·5	55·5	55·2	54·8	54·5
605	4·3	22 06	30·8	30·7	30·9	31·4	32·2	33·1	34·1	34·8	35·2	35·1	34·8	34·4	34·1
606	4·0	22 07	04·4	04·3	04·4	04·8	05·7	06·7	07·6	08·3	08·5	08·4	08·1	07·6	07·2
607	2·2	22 08	19·1	18·9	19·2	19·8	20·8	22·2	23·4	24·4	24·9	24·8	24·3	23·6	23·1
608	4·4	22 10	02·7	02·5	02·6	03·0	03·9	04·9	05·9	06·6	06·9	06·8	06·3	05·8	05·3
609	3·7	22 10	16·2	16·1	16·3	16·7	17·5	18·4	19·3	20·0	20·3	20·3	20·0	19·6	19·3
610	3·6	22 10	53·2	52·6	52·6	53·2	54·4	55·8	57·1	58·0	58·3	57·9	57·1	56·0	55·1
611	4·3	22 16	54·5	54·4	54·6	55·0	55·8	56·7	57·7	58·4	58·8	58·7	58·5	58·1	57·8
612	2·9	22 18	35·3	35·0	35·2	36·0	37·3	38·9	40·6	41·9	42·5	42·4	41·6	40·6	39·8
613	4·0	22 21	43·8	43·7	43·8	44·2	45·0	45·9	46·8	47·5	47·9	47·9	47·6	47·2	47·0
614	3·7	22 28	54·3	54·2	54·3	54·7	55·4	56·3	57·3	58·0	58·4	58·4	58·1	57·7	57·4
615	4·0	22 29	21·1	21·0	21·1	21·6	22·5	23·7	25·0	25·9	26·5	26·5	26·1	25·5	25·0
616	4·0	22 29	12·6	11·9	11·8	12·3	13·4	14·9	16·3	17·2	17·6	17·4	16·6	15·6	14·6
617	4·4	22 31	35·2	35·0	35·2	35·6	36·4	37·5	38·6	39·5	39·9	40·0	39·6	39·2	38·8
618	3·8	22 31	20·5	20·0	19·9	20·4	21·4	22·6	23·8	24·7	25·1	24·9	24·4	23·6	22·8
619	4·1	22 35	25·8	25·7	25·8	26·2	26·9	27·8	28·7	29·5	29·9	29·9	29·7	29·3	29·0
620	4·2	22 40	44·1	44·0	44·1	44·5	45·2	46·3	47·3	48·1	48·6	48·7	48·4	48·0	47·6
621	3·6	22 41	32·1	31·9	31·9	32·3	33·0	33·9	34·9	35·6	36·0	36·0	35·8	35·4	35·1
622	2·2	22 42	45·1	44·8	44·9	45·3	46·3	47·5	48·8	49·9	50·5	50·5	50·1	49·5	48·9
623	3·1	22 43	04·1	03·8	03·8	04·1	04·9	05·9	06·9	07·7	08·1	08·1	07·8	07·3	06·9
624	4·3	22 46	08·9	06·6	06·3	07·9	11·2	15·8	20·7	24·6	26·6	26·3	23·7	19·9	16·1
625	4·1	22 46	36·0	35·8	35·8	36·1	36·8	37·8	38·8	39·5	40·0	40·0	39·7	39·3	38·9
626	3·7	22 48	38·3	38·0	38·0	38·5	39·5	40·8	42·2	43·3	44·0	44·0	43·6	42·9	42·2
627	4·2	22 49	40·2	40·0	40·1	40·4	41·1	42·0	43·0	43·8	44·3	44·4	44·1	43·8	43·5
628	3·7	22 49	42·9	41·8	41·4	41·8	43·2	44·9	46·7	47·9	48·5	48·3	47·4	46·0	44·6
629	3·7	22 50	04·4	04·1	04·1	04·4	05·1	06·1	07·1	07·9	08·3	08·3	08·1	07·6	07·2
630	3·8	22 52	41·5	41·3	41·4	41·7	42·4	43·3	44·2	45·0	45·5	45·6	45·4	45·0	44·7
631	3·5	22 54	43·7	43·5	43·6	43·9	44·6	45·5	46·5	47·3	47·8	47·9	47·7	47·3	47·0
632 d	1·3	22 57	43·9	43·7	43·8	44·1	44·8	45·9	46·9	47·8	48·4	48·5	48·2	47·8	47·4
633	4·2	23 00	57·8	57·4	57·3	57·8	58·7	60·0	61·4	62·6	63·4	63·5	63·0	62·3	61·6
634	3·6	23 01	59·1	58·7	58·6	58·9	59·7	60·8	61·9	62·8	63·3	63·3	63·0	62·4	61·8
635	2·6	23 03	50·7	50·4	50·3	50·6	51·3	52·3	53·3	54·1	54·6	54·7	54·5	54·0	53·6
636	2·6	23 04	50·1	49·9	49·8	50·1	50·8	51·7	52·6	53·4	53·9	54·0	53·8	53·5	53·1
637	3·8	23 09	31·6	31·4	31·4	31·7	32·3	33·3	34·3	35·2	35·7	35·8	35·7	35·3	34·9
638	4·1	23 10	26·5	26·1	26·1	26·4	27·2	28·4	29·7	30·8	31·4	31·6	31·3	30·7	30·1
639	4·4	23 14	24·1	23·9	23·8	24·1	24·7	25·6	26·6	27·4	27·9	28·1	27·9	27·6	27·3
640	3·8	23 17	14·7	14·5	14·4	14·7	15·3	16·2	17·1	18·0	18·5	18·6	18·5	18·2	17·9
641	4·1	23 17	30·7	30·1	30·0	30·3	31·3	32·7	34·3	35·7	36·6	36·7	36·3	35·4	34·6
642	4·2	23 23	03·0	02·8	02·7	03·0	03·6	04·5	05·5	06·4	07·0	07·2	07·0	06·7	06·3
643	4·4	23 28	02·8	02·5	02·5	02·7	03·2	04·1	05·1	05·9	06·4	06·6	06·5	06·2	05·9
644	4·5	23 33	03·1	02·8	02·7	02·9	03·6	04·6	05·7	06·8	07·5	07·7	07·5	07·1	06·6
645	4·0	23 37	38·3	37·7	37·4	37·6	38·3	39·4	40·6	41·6	42·3	42·5	42·3	41·7	41·1
646	4·3	23 38	12·7	12·1	11·9	12·0	12·7	13·8	15·0	16·0	16·6	16·8	16·6	16·1	15·5
647	3·4	23 39	23·4	20·7	19·2	19·3	21·1	24·0	27·2	29·8	31·4	31·5	30·3	27·9	25·1
648	4·3	23 40	01·9	01·6	01·5	01·7	02·2	03·1	04·1	04·9	05·5	05·7	05·6	05·3	05·0
649	4·3	23 40	29·0	28·5	28·2	28·3	29·0	30·1	31·3	32·3	33·0	33·2	33·0	32·5	31·9
650	4·0	23 59	23·6	23·3	23·2	23·3	23·8	24·6	25·6	26·5	27·1	27·4	27·3	27·1	26·8

The figures given refer to the beginning of the month, and should be interpolated to the actual date by means of the table on page 73.

No.		Name	Dec	Jan.	F.	M.	Apr.	M.	J.	July	A.	S.	Oct.	N.	D.	Jan.
			° ′	″	″	″	″	″	″	″	″	″	″	″	″	″
601	δ	Capricorni	S 16 06	77	77	75	72	68	62	58	56	55	56	58	60	61
602	π²	Cygni	N 49 18	73	65	56	50	48	52	59	69	79	88	93	93	89
603	γ	Gruis	S 37 21	36	32	27	21	15	09	07	08	11	15	19	21	20
604	α	Aquarii	S 0 18	44	46	48	47	44	39	33	28	24	23	23	25	27
605	ι	Aquarii	S 13 51	47	47	46	43	39	33	28	25	24	25	27	29	30
606	ι	Pegasi	N 25 21	18	12	06	03	03	07	14	22	30	36	38	38	34
607	α	Gruis	S 46 56	83	79	72	64	57	52	50	51	56	62	67	69	67
608	π	Pegasi	N 33 11	19	13	06	01	01	04	11	20	29	36	39	40	36
609	θ	Pegasi	N 6 12	22	19	16	16	18	23	30	36	40	43	43	42	39
610	ζ	Cephei	N 58 12	48	40	31	23	20	22	29	39	50	60	66	68	65
611	θ	Aquarii	S 7 46	34	35	35	33	29	23	18	14	12	11	13	15	16
612	α	Tucanae	S 60 14	80	73	65	56	47	42	41	44	50	58	63	66	63
613	γ	Aquarii	S 1 22	46	48	49	48	45	40	34	29	25	24	24	26	28
614	ζ	Aquarii	S 0 00	43	46	47	46	43	38	32	27	23	21	22	23	25
615	δ¹	Gruis	S 43 28	87	82	76	69	61	55	52	52	56	62	67	70	69
616	δ	Cephei	N 58 25	40	33	24	16	12	14	20	30	41	51	58	60	58
617	β	Piscis Aust.	S 32 19	86	83	79	73	66	60	56	55	57	61	65	68	68
618	α	Lacertae	N 50 17	41	34	26	19	16	18	24	34	44	53	60	61	59
619	η	Aquarii	S 0 06	33	36	37	36	33	28	22	17	13	11	11	13	15
620	ε	Piscis Aust.	S 27 01	75	74	70	65	58	52	47	46	47	50	53	56	57
621	ζ	Pegasi	N 10 50	26	23	20	18	20	25	31	38	43	47	48	47	45
622	β	Gruis	S 46 52	48	43	36	28	20	13	10	11	15	21	27	30	29
623	η	Pegasi	N 30 13	56	51	45	40	39	43	49	57	66	72	77	77	74
624	β	Octantis	S 81 21	101	92	82	71	61	56	55	59	67	76	83	86	82
625	λ	Pegasi	N 23 34	34	29	24	21	21	25	31	39	46	52	55	55	53
626	ε	Gruis	S 51 18	44	39	32	23	15	08	05	06	11	17	24	27	26
627	τ	Aquarii	S 13 34	67	68	67	64	59	53	47	43	42	43	45	47	49
628	ι	Cephei	N 66 12	50	43	34	26	21	21	26	35	46	57	66	69	68
629	μ	Pegasi	N 24 36	44	39	34	30	30	34	40	48	56	62	65	65	63
630	λ	Aquarii	S 7 33	78	80	80	78	73	67	62	57	55	54	56	58	60
631	δ	Aquarii	S 15 48	49	49	48	45	39	33	28	24	23	24	26	29	31
632		Fomalhaut (α PsA)	S 29 36	59	57	53	47	40	34	29	27	28	32	36	39	40
633	ζ	Gruis	S 52 44	59	54	46	37	28	21	18	19	23	30	37	40	40
634	ο	Andromedae	N 42 19	78	72	65	59	56	58	64	72	82	90	96	99	97
635	β	Pegasi	N 28 05	39	34	28	24	23	26	33	40	49	55	59	60	58
636	α	Pegasi	N 15 12	55	51	48	45	47	51	57	64	70	75	77	76	74
637	88	Aquarii	S 21 09	56	56	53	49	43	36	30	27	27	29	32	35	37
638	ι	Gruis	S 45 13	90	87	80	72	63	56	52	52	55	61	67	71	71
639	φ	Aquarii	S 6 02	27	29	29	27	23	17	11	06	04	03	04	06	08
640	γ	Piscium	N 3 17	29	26	25	25	28	33	39	45	49	52	52	51	48
641	γ	Tucanae	S 58 13	53	48	40	30	20	13	09	10	15	23	30	34	33
642	98	Aquarii	S 20 05	37	37	35	30	24	17	12	08	08	09	13	16	18
643	θ	Piscium	N 6 23	18	15	13	13	15	20	26	32	37	40	41	40	38
644	β	Sculptoris	S 37 48	46	44	39	32	23	16	10	09	11	15	21	25	26
645	λ	Andromedae	N 46 27	75	71	64	57	53	53	58	66	76	85	92	96	95
646	ι	Andromedae	N 43 16	51	46	40	33	30	31	36	44	53	62	69	72	71
647	γ	Cephei	N 77 38	50	46	37	28	21	19	22	29	40	51	62	68	70
648	ι	Piscium	N 5 38	08	05	03	03	05	10	16	22	27	30	30	29	27
649	κ	Andromedae	N 44 20	49	44	37	31	27	28	33	40	50	59	66	69	69
650	ω	Piscium	N 6 52	22	20	18	17	19	24	30	36	41	44	45	44	43

No.	Mag.	RA	Jan.	Feb.	Mar.	Apr.	May	June	July	Aug.	Sept.	Oct.	Nov.	Dec.	Jan.
		h m	s	s	s	s	s	s	s	s	s	s	s	s	s
651	5·4	0 47	53·2	50·9	49·2	48·6	49·5	51·6	54·2	56·8	58·9	60·0	59·9	58·7	56·7
652	5·2	2 05	43·3	40·8	38·5	37·1	37·1	38·7	41·4	44·4	47·4	49·5	50·5	50·1	48·6
653	5·5	3 20	37·1	34·7	31·9	29·6	28·8	29·7	32·1	35·5	39·2	42·4	44·6	45·3	44·5
654	5·2	4 10	26·9	24·4	20·7	17·3	15·3	15·7	18·1	22·1	27·0	31·6	35·3	37·2	36·9
655	5·1	5 00	47·8	46·0	42·4	38·5	35·8	35·3	37·1	40·7	45·7	51·0	55·6	58·5	59·2
656	5·2	5 22	57·1	56·0	53·2	50·0	47·7	47·0	48·2	51·0	55·1	59·5	63·4	66·2	67·2
657	4·7	7 00	24·6	24·9	23·5	21·1	18·7	17·2	17·1	18·5	21·1	24·6	28·3	31·6	33·7
658	5·3	8 05	09·4	10·8	09·9	07·4	04·2	01·8	00·7	01·3	03·6	07·3	11·7	16·1	19·4
659	4·6	9 37	25·3	28·6	29·1	27·3	23·8	20·2	17·5	16·4	17·3	20·2	24·7	29·9	34·7
660	5·0	10 35	16·6	19·2	20·3	19·8	18·1	16·0	14·1	12·9	12·8	13·9	16·2	19·2	22·4
661	5·1	12 12	17·1	20·7	22·9	23·6	22·5	20·4	17·9	15·6	14·1	13·8	14·9	17·5	20·8
662	5·3	12 48	66·9	73·3	77·9	79·8	78·5	74·8	69·8	64·8	60·8	59·1	60·0	63·6	69·1
663	5·0	14 08	48·1	51·5	54·5	56·5	56·9	55·8	53·6	50·8	48·0	46·0	45·4	46·3	48·7
375	4·4	14 27	28·8	31·7	34·5	36·4	36·9	36·2	34·4	32·0	29·5	27·6	26·8	27·4	29·3
389	2·2	14 50	39·3	41·9	44·4	46·3	47·1	46·7	45·2	43·1	40·7	38·8	37·8	38·1	39·7
417	4·3	15 43	55·3	58·1	61·3	64·1	65·7	65·6	64·1	61·4	58·2	55·1	53·1	52·5	53·7
664	5·0	16 17	22·9	25·0	27·8	30·3	32·0	32·4	31·4	29·3	26·5	23·7	21·6	20·8	21·5
453	4·4	16 45	39·3	42·4	46·9	51·5	54·7	55·6	54·1	50·5	45·4	40·1	35·7	33·3	33·6
665	5·0	17 49	16·5	17·8	20·3	23·2	25·7	27·1	27·1	25·5	22·7	19·5	16·4	14·3	13·8
666	5·1	19 08	60·2	60·4	62·1	64·8	67·6	69·7	70·5	69·9	67·9	65·0	61·8	59·1	57·7
557	4·4	20 08	43·8	43·2	44·3	46·8	49·9	52·6	54·2	54·3	52·8	50·0	46·6	43·4	41·2
667	5·2	20 31	24·3	23·6	24·3	26·2	28·8	31·2	32·8	33·2	32·2	30·2	27·6	24·9	23·0
668	4·8	22 47	23·1	18·5	16·7	17·9	21·9	27·2	32·2	35·7	37·0	35·7	32·0	26·7	21·2
669	4·5	23 07	55·1	53·0	52·0	52·5	54·3	56·9	59·6	61·6	62·6	62·3	60·9	58·7	56·3
647	3·4	23 39	23·4	20·7	19·2	19·3	21·1	24·0	27·2	29·8	31·4	31·5	30·3	27·9	25·1
670	4·7	0 01	39·8	37·6	36·4	36·4	37·7	40·3	43·5	46·5	48·7	49·4	48·6	46·4	43·8
7	2·9	0 25	50·1	47·8	46·3	46·0	47·1	49·5	52·6	55·9	58·3	59·4	58·8	56·8	54·2
671	4·7	2 50	31·2	28·7	26·3	24·3	23·4	23·7	25·3	27·8	30·5	32·6	33·6	33·1	31·4
91	3·2	3 47	16·7	14·5	12·0	09·8	08·2	07·9	08·9	10·9	13·4	15·7	17·2	17·4	16·2
672	5·1	6 10	16·7	15·4	13·1	10·4	07·8	06·0	05·4	06·2	08·1	10·6	13·0	14·6	14·8
673	4·1	8 18	34·6	34·7	33·2	30·7	27·6	24·7	22·7	21·9	22·8	24·9	27·8	30·6	32·4
227	4·3	8 20	41·1	41·2	39·7	37·0	33·8	30·8	28·6	27·8	28·6	30·8	33·8	36·7	38·6
674	5·2	9 33	56·4	57·8	57·1	54·5	50·7	46·4	42·6	40·3	40·0	42·0	45·6	49·9	53·4
289	4·1	10 35	32·7	34·8	35·2	34·2	31·9	28·8	25·7	23·3	22·3	23·1	25·6	29·0	32·4
675	4·6	10 45	51·5	54·2	54·8	53·7	51·1	47·4	43·6	40·6	39·2	40·0	42·7	46·8	50·8
676	4·9	11 59	43·5	46·6	48·3	48·6	47·6	45·4	42·6	39·8	37·9	37·6	39·0	42·0	45·6
677	5·0	12 04	52·6	55·4	57·0	57·4	56·6	54·8	52·3	50·0	48·3	48·0	49·2	51·8	55·0
325	4·4	12 18	27·3	30·8	33·0	33·6	32·8	30·6	27·6	24·5	22·2	21·5	22·8	25·9	29·9
678	5·0	13 25	14·2	17·2	19·4	20·8	21·1	20·3	18·7	16·5	14·6	13·5	13·8	15·6	18·4
679	4·9	14 18	23·0	27·9	32·2	35·3	36·8	36·4	34·1	30·6	26·9	24·3	23·8	25·8	29·9
680	4·3	14 27	06·1	13·0	19·1	23·6	26·0	25·6	22·4	17·3	11·9	08·0	07·0	09·5	15·1
386	3·8	14 48	00·4	04·6	08·3	11·3	13·1	13·3	11·9	09·2	06·2	03·8	03·0	04·3	07·5
681	5·0	14 58	01·1	04·6	07·8	10·3	12·0	12·4	11·3	09·3	06·8	04·8	04·0	05·0	07·6
682	4·8	16 20	30·1	33·7	37·7	41·5	44·5	46·2	46·2	44·4	41·6	38·8	36·9	36·9	39·0
443	3·9	16 33	36·4	40·0	44·1	48·0	51·2	53·1	53·2	51·6	48·8	45·9	43·9	43·6	45·6
683	4·2	16 43	13·6	16·8	20·4	23·9	27·0	28·8	29·1	27·8	25·4	22·7	20·9	20·5	22·2
684	5·2	21 04	49·9	49·9	51·1	53·5	56·8	60·4	63·4	65·3	65·5	64·2	61·9	59·4	57·8
596	3·7	21 41	35·0	34·4	35·2	37·2	40·3	43·9	47·2	49·5	50·2	49·2	47·0	44·4	42·4
685	5·1	22 20	07·6	06·1	06·3	08·2	11·7	16·1	20·5	23·8	25·3	24·5	22·0	18·4	15·3
624	4·3	22 46	08·9	06·6	06·3	07·9	11·2	15·8	20·7	24·6	26·6	26·3	23·7	19·9	16·1

The figures given refer to the beginning of the month, and should be interpolated to the actual date by means of the table on page 73.

No.	Name		Dec	Jan.	F.	M.	Apr.	M.	J.	July	A.	S.	Oct.	N.	D.	Jan.
651	23	Cassiopeiae	N 74 51	44	42	36	27	19	15	15	20	29	40	51	60	64
652	49	Cassiopeiae	N 76 07	41	43	40	33	24	18	15	17	23	32	43	53	60
653	BS 0961	(Cephei)	N 77 44	41	47	46	41	33	25	20	19	21	28	38	48	57
654	BS 1230	(Cephei)	N 80 42	22	29	31	28	20	12	05	02	02	06	15	25	34
655	BS 1523	(Cephei)	N 81 11	55	64	68	66	60	52	44	39	36	39	45	54	63
656	BS 1686	(Cam)	N 79 13	64	73	77	77	71	64	56	50	47	48	53	61	70
657	BS 2527	(Cam)	N 76 58	27	37	44	47	46	40	32	24	17	13	13	17	25
658	BS 3082	(Cam)	N 79 28	21	31	39	45	45	41	34	25	16	09	06	07	13
659	BS 3751	(Draconis)	N 81 18	53	60	70	77	82	81	75	66	56	46	39	36	38
660	BS 4126	(Draconis)	N 75 41	58	62	71	79	85	87	84	77	66	56	46	40	40
661	BS 4646	(Cam)	N 77 35	65	66	72	81	89	95	95	90	81	70	58	49	44
662*	BS 4893	(Cam)	N 83 23	71	71	76	85	94	00	102	98	90	79	67	56	50
663	4	Ursae Minoris	N 77 31	64	60	62	69	78	87	91	91	87	77	66	54	45
375	5	Ursae Minoris	N 75 40	61	56	57	64	73	82	87	88	84	75	64	52	43
389	β	Ursae Minoris	N 74 08	38	32	33	39	48	57	63	65	62	54	43	31	21
417	ζ	Ursae Minoris	N 77 46	67	60	58	62	71	80	88	92	91	85	76	64	53
664	η	Ursae Minoris	N 75 44	53	44	41	44	52	62	70	76	77	73	64	53	42
453	ε	Ursae Minoris	N 82 01	54	45	41	43	50	59	68	74	76	73	66	56	45
665	35	Draconis	N 76 57	41	30	24	23	28	37	47	55	61	61	57	48	37
666	59	Draconis	N 76 33	50	39	30	27	29	36	46	56	64	68	68	62	53
557	κ	Cephei	N 77 42	67	56	47	41	41	46	55	66	76	83	85	82	75
667	73	Draconis	N 74 57	47	36	27	20	19	24	33	43	54	62	65	63	57
668	BS 8702	(Cephei)	N 83 09	64	58	49	40	34	33	37	46	57	69	78	83	83
669	π	Cephei	N 75 23	66	61	52	43	37	36	40	48	59	71	80	86	86
647	γ	Cephei	N 77 38	50	46	37	28	21	19	22	29	40	51	62	68	70
670	θ	Octantis	S 77 02	104	98	89	78	66	58	54	55	61	69	78	84	84
7	β	Hydri	S 77 14	62	57	48	37	26	16	11	11	17	25	34	40	41
671	ν	Hydri	S 75 03	53	54	51	43	32	21	11	05	05	10	19	29	36
91	γ	Hydri	S 74 13	75	80	79	73	64	52	42	34	32	35	43	53	62
672	α	Mensae	S 74 44	76	85	90	91	88	80	70	60	52	50	54	63	74
673	α	Chamaeleontis	S 76 55	21	32	42	49	52	49	42	33	23	17	15	20	30
227	θ	Chamaeleontis	S 77 29	15	26	36	43	45	43	36	27	17	11	09	14	24
674	ζ	Chamaeleontis	S 80 56	42	53	64	73	79	81	77	69	60	51	47	49	56
289	γ	Chamaeleontis	S 78 36	42	52	63	74	82	86	85	79	70	61	55	54	60
675	δ²	Chamaeleontis	S 80 32	38	48	59	70	79	84	83	78	69	60	53	52	57
676	ε	Chamaeleontis	S 78 13	32	40	50	62	72	79	82	79	72	63	55	51	53
677	κ	Chamaeleontis	S 76 31	22	30	40	51	62	69	71	69	62	53	45	41	43
325	β	Chamaeleontis	S 79 18	57	64	74	85	96	04	107	105	98	90	81	77	78
678	ι¹	Muscae	S 74 53	28	33	41	51	62	71	76	77	72	65	56	50	49
679	η	Apodis	S 81 00	37	38	44	53	64	74	82	85	83	76	67	59	55
680	δ	Octantis	S 83 40	13	14	19	28	39	50	57	61	59	53	44	36	31
386	α	Apodis	S 79 02	49	50	54	63	73	83	91	95	94	88	80	72	67
681	R	Apodis	S 76 39	53	53	58	66	75	85	93	97	97	91	83	75	70
682*	δ¹	Apodis	S 78 41	48	44	45	49	57	66	75	82	84	82	75	67	59
443	γ	Apodis	S 78 53	52	48	48	51	59	68	77	84	87	85	79	71	63
683	β	Apodis	S 77 31	05	01	00	04	10	19	28	35	39	37	32	23	15
684	α	Octantis	S 77 00	75	65	56	47	42	40	44	51	59	67	70	69	63
596	ν	Octantis	S 77 22	72	63	53	43	36	33	35	41	50	58	63	63	57
685	ε	Octantis	S 80 25	70	61	51	40	32	27	28	33	41	50	56	57	53
624	β	Octantis	S 81 21	101	92	82	71	61	56	55	59	67	76	83	86	82

* No., mag., dist. and p.a. of companion star: **662**, 5·8, 22″, 326°   **682**, 5·2, 103″, 11°

Name	BS 0285 (Cephei)		α Ursae Min. (*Polaris*)		BS 2609 (Cephei)		δ Ursae Min.		BS 8546 (Cephei)		GST
Mag.	4·5		2·1		5·3		4·4		5·4		
Date	RA	Dec	RA	Dec	RA	Dec	RA	Dec	RA	Dec	
0^h UT	1^h	N 86°	2^h	N 89°	7^h	N 87°	17^h	N 86°	22^h	N 86°	
	m s	′ ″	m s	′ ″	m s	′ ″	m s	′ ″	m s	′ ″	h m
Jan. 1	09 03	16 18	34 35	16 35	41 40	00 52	31 19	35 01	12 49	07 15	06 42
11	09 00	16 19	34 20	16 37	41 42	00 56	31 20	34 58	12 46	07 13	07 21
21	08 56	16 19	34 03	16 38	41 42	00 59	31 21	34 54	12 44	07 11	08 01
31	08 53	16 18	33 46	16 39	41 43	01 02	31 23	34 52	12 42	07 08	08 40
Feb. 10	08 50	16 17	33 29	16 39	41 42	01 05	31 26	34 49	12 41	07 05	09 20
20	08 47	16 15	33 12	16 38	41 40	01 08	31 29	34 48	12 40	07 02	09 59
Mar. 2	08 45	16 13	32 56	16 37	41 38	01 11	31 33	34 46	12 40	06 59	10 38
12	08 43	16 10	32 42	16 35	41 35	01 13	31 36	34 46	12 40	06 56	11 18
22	08 42	16 07	32 31	16 33	41 31	01 15	31 40	34 46	12 42	06 53	11 57
Apr. 1	08 41	16 04	32 22	16 30	41 27	01 16	31 43	34 46	12 43	06 50	12 37
11	08 41	16 01	32 16	16 28	41 24	01 16	31 47	34 48	12 45	06 48	13 16
21	08 42	15 58	32 14	16 24	41 20	01 16	31 49	34 50	12 48	06 46	13 56
May 1	08 43	15 55	32 15	16 21	41 16	01 15	31 52	34 52	12 51	06 45	14 35
11	08 45	15 53	32 17	16 18	41 13	01 14	31 54	34 54	12 54	06 44	15 14
21	08 47	15 51	32 24	16 16	41 10	01 12	31 55	34 57	12 57	06 44	15 54
31	08 50	15 49	32 35	16 13	41 07	01 10	31 55	35 00	13 00	06 45	16 33
June 10	08 53	15 48	32 46	16 11	41 05	01 07	31 55	35 04	13 03	06 46	17 13
20	08 56	15 48	32 59	16 10	41 04	01 05	31 55	35 07	13 05	06 48	17 52
30	09 00	15 48	33 15	16 09	41 04	01 02	31 53	35 09	13 08	06 50	18 32
July 10	09 03	15 48	33 32	16 08	41 04	00 58	31 51	35 12	13 10	06 53	19 11
20	09 06	15 49	33 49	16 08	41 05	00 55	31 49	35 15	13 11	06 56	19 50
30	09 09	15 51	34 07	16 08	41 07	00 52	31 46	35 17	13 12	06 59	20 30
Aug. 9	09 13	15 53	34 26	16 09	41 09	00 49	31 43	35 19	13 13	07 03	21 09
19	09 15	15 56	34 43	16 11	41 12	00 46	31 39	35 20	13 13	07 07	21 49
29	09 18	15 59	34 59	16 13	41 15	00 44	31 35	35 21	13 13	07 10	22 28
Sept. 8	09 20	16 02	35 15	16 15	41 19	00 41	31 31	35 21	13 12	07 14	23 08
18	09 21	16 06	35 30	16 18	41 23	00 39	31 27	35 21	13 11	07 18	23 47
28	09 22	16 09	35 42	16 21	41 27	00 38	31 22	35 21	13 09	07 21	00 26
Oct. 8	09 23	16 13	35 52	16 24	41 32	00 37	31 18	35 20	13 07	07 24	01 06
18	09 23	16 17	36 01	16 28	41 38	00 36	31 14	35 18	13 05	07 27	01 45
28	09 23	16 21	36 07	16 31	41 42	00 35	31 11	35 16	13 02	07 30	02 25
Nov. 7	09 22	16 25	36 10	16 35	41 47	00 36	31 07	35 14	12 59	07 32	03 04
17	09 21	16 28	36 09	16 39	41 52	00 37	31 04	35 11	12 55	07 33	03 44
27	09 19	16 31	36 07	16 43	41 57	00 38	31 02	35 08	12 52	07 34	04 23
Dec. 7	09 17	16 34	36 02	16 46	42 01	00 40	31 00	35 05	12 49	07 34	05 02
17	09 14	16 36	35 52	16 49	42 04	00 42	31 00	35 01	12 45	07 34	05 42
27	09 11	16 38	35 40	16 52	42 07	00 45	30 59	34 57	12 42	07 33	06 21
37	09 08	16 39	35 28	16 54	42 10	00 48	31 00	34 54	12 39	07 32	07 01

Name	ζ Octantis		ι Octantis		χ Octantis		σ Octantis		τ Octantis		
Mag.	5·4		5·4		5·2		5·5		5·6		
Date	RA	Dec	RA	Dec	RA	Dec	RA	Dec	RA	Dec	GST
0ʰ UT	8ʰ	S 85°	12ʰ	S 85°	18ʰ	S 87°	21ʰ	S 88°	23ʰ	S 87°	
	m s	′ ″	m s	′ ″	m s	′ ″	m s	′ ″	m s	′ ″	h m
Jan. 1	56 40	39 59	55 09	07 36	55 11	36 15	09 15	57 12	28 07	28 43	06 42
11	56 41	40 02	55 12	07 37	55 12	36 12	09 11	57 08	28 02	28 41	07 21
21	56 41	40 06	55 14	07 39	55 15	36 09	09 09	57 05	27 59	28 38	08 01
31	56 41	40 10	55 17	07 41	55 18	36 06	09 10	57 01	27 56	28 35	08 40
Feb. 10	56 40	40 13	55 19	07 44	55 23	36 03	09 12	56 57	27 54	28 32	09 20
20	56 39	40 17	55 21	07 47	55 27	36 01	09 16	56 54	27 52	28 28	09 59
Mar. 2	56 38	40 20	55 22	07 50	55 33	35 59	09 23	56 51	27 52	28 25	10 38
12	56 36	40 24	55 24	07 54	55 39	35 57	09 31	56 47	27 51	28 21	11 18
22	56 33	40 26	55 25	07 57	55 45	35 56	09 40	56 44	27 52	28 17	11 57
Apr. 1	56 31	40 29	55 25	08 01	55 51	35 56	09 51	56 42	27 54	28 13	12 37
11	56 28	40 31	55 25	08 05	55 57	35 56	10 03	56 39	27 56	28 09	13 16
21	56 25	40 32	55 25	08 09	56 03	35 56	10 15	56 38	27 58	28 06	13 56
May 1	56 22	40 33	55 24	08 12	56 09	35 57	10 28	56 36	28 02	28 03	14 35
11	56 18	40 34	55 23	08 16	56 15	35 58	10 41	56 36	28 06	28 00	15 14
21	56 15	40 34	55 22	08 19	56 20	36 00	10 54	56 35	28 10	27 58	15 54
31	56 12	40 33	55 20	08 22	56 24	36 02	11 07	56 35	28 15	27 56	16 33
June 10	56 10	40 32	55 18	08 24	56 28	36 05	11 19	56 36	28 20	27 55	17 13
20	56 07	40 30	55 16	08 25	56 31	36 07	11 30	56 38	28 25	27 54	17 52
30	56 05	40 28	55 14	08 27	56 33	36 10	11 41	56 39	28 30	27 54	18 32
July 10	56 03	40 26	55 12	08 27	56 35	36 13	11 49	56 41	28 35	27 54	19 11
20	56 02	40 23	55 09	08 27	56 35	36 16	11 55	56 44	28 39	27 55	19 50
30	56 01	40 20	55 07	08 27	56 34	36 20	12 01	56 47	28 44	27 57	20 30
Aug. 9	56 00	40 17	55 05	08 26	56 33	36 22	12 04	56 50	28 47	27 59	21 09
19	56 00	40 14	55 03	08 24	56 30	36 25	12 05	56 53	28 50	28 01	21 49
29	56 01	40 11	55 01	08 22	56 27	36 28	12 03	56 56	28 53	28 04	22 28
Sept. 8	56 02	40 08	54 59	08 19	56 24	36 29	12 01	56 59	28 54	28 07	23 08
18	56 03	40 05	54 59	08 17	56 19	36 31	11 55	57 02	28 54	28 10	23 47
28	56 06	40 03	54 58	08 13	56 14	36 32	11 47	57 04	28 54	28 13	00 26
Oct. 8	56 08	40 01	54 58	08 10	56 09	36 32	11 39	57 06	28 53	28 16	01 06
18	56 10	40 00	54 58	08 08	56 05	36 32	11 29	57 07	28 51	28 19	01 45
28	56 13	40 00	54 59	08 05	56 00	36 30	11 18	57 08	28 47	28 21	02 25
Nov. 7	56 16	40 00	55 00	08 02	55 56	36 29	11 06	57 08	28 44	28 23	03 04
17	56 19	40 00	55 02	08 00	55 52	36 27	10 55	57 08	28 40	28 24	03 44
27	56 21	40 02	55 04	07 58	55 49	36 24	10 45	57 07	28 35	28 25	04 23
Dec. 7	56 24	40 04	55 07	07 57	55 47	36 21	10 35	57 05	28 30	28 25	05 02
17	56 26	40 06	55 09	07 56	55 46	36 18	10 26	57 03	28 25	28 25	05 42
27	56 27	40 09	55 12	07 56	55 46	36 15	10 20	57 00	28 21	28 24	06 21
37	56 29	40 13	55 15	07 57	55 48	36 12	10 15	56 57	28 17	28 22	07 01

LST	0^h		1^h		2^h		3^h		4^h		5^h	
	a_0	b_0	a_0	b_0	a_0	b_0	a_0	b_0	a_0	b_0	a_0	b_0
m	′	′	′	′	′	′	′	′	′	′	′	′
0	−33·9	+27·4	−39·8	+17·6	−43·0	+6·6	−43·2	−5·0	−40·4	−16·1	−34·9	−26·2
3	34·2	27·0	40·1	17·1	43·1	6·0	43·2	5·5	40·2	16·7	34·5	26·6
6	34·6	26·5	40·3	16·6	43·2	5·4	43·1	6·1	40·0	17·2	34·2	27·1
9	34·9	26·1	40·5	16·0	43·2	4·8	43·0	6·7	39·8	17·7	33·8	27·5
12	35·3	25·6	40·7	15·5	43·3	4·3	42·9	7·2	39·5	18·3	33·5	28·0
15	−35·6	+25·1	−40·9	+14·9	−43·3	+3·7	−42·8	−7·8	−39·3	−18·8	−33·1	−28·4
18	35·9	24·7	41·1	14·4	43·4	3·1	42·7	8·4	39·1	19·3	32·7	28·8
21	36·3	24·2	41·3	13·9	43·4	2·5	42·6	9·0	38·8	19·8	32·3	29·3
24	36·6	23·7	41·4	13·3	43·5	2·0	42·5	9·5	38·5	20·3	31·9	29·7
27	36·9	23·2	41·6	12·8	43·5	1·4	42·3	10·1	38·3	20·8	31·6	30·1
30	−37·2	+22·7	−41·8	+12·2	−43·5	+0·8	−42·2	−10·6	−38·0	−21·3	−31·2	−30·5
33	37·5	22·2	41·9	11·6	43·5	+0·2	42·1	11·2	37·7	21·8	30·8	30·9
36	37·8	21·7	42·1	11·1	43·5	−0·3	41·9	11·8	37·4	22·3	30·3	31·3
39	38·0	21·2	42·2	10·5	43·5	0·9	41·7	12·3	37·1	22·8	29·9	31·7
42	38·3	20·7	42·4	10·0	43·5	1·5	41·6	12·9	36·8	23·3	29·5	32·1
45	−38·6	+20·2	−42·5	+9·4	−43·5	−2·1	−41·4	−13·4	−36·5	−23·8	−29·1	−32·5
48	38·8	19·7	42·6	8·8	43·4	2·7	41·2	14·0	36·2	24·3	28·7	32·9
51	39·1	19·2	42·7	8·3	43·4	3·2	41·0	14·5	35·9	24·8	28·2	33·3
54	39·3	18·7	42·8	7·7	43·3	3·8	40·9	15·1	35·5	25·2	27·8	33·6
57	39·6	18·1	42·9	7·1	43·3	4·4	40·7	15·6	35·2	25·7	27·4	34·0
60	−39·8	+17·6	−43·0	+6·6	−43·2	−5·0	−40·4	−16·1	−34·9	−26·2	−26·9	−34·3

Lat.	a_1	b_1	a_1	b_1	a_1	b_1	a_1	b_1	a_1	b_1	a_1	b_1
°												
0	−·1	−·3	·0	−·2	·0	·0	·0	+·2	−·1	+·3	−·2	+·3
10	−·1	−·2	·0	−·1	·0	·0	·0	+·1	−·1	+·2	−·1	+·3
20	−·1	−·2	·0	−·1	·0	·0	·0	+·1	−·1	+·2	−·1	+·2
30	·0	−·1	·0	−·1	·0	·0	·0	+·1	·0	+·1	−·1	+·2
40	·0	−·1	·0	−·1	·0	·0	·0	·0	·0	+·1	·0	+·1
45	·0	·0	·0	·0	·0	·0	·0	·0	·0	·0	·0	+·1
50	·0	·0	·0	·0	·0	·0	·0	·0	·0	·0	·0	·0
55	·0	+·1	·0	·0	·0	·0	·0	·0	·0	−·1	·0	−·1
60	·0	+·1	·0	+·1	·0	·0	·0	−·1	·0	−·1	+·1	−·1
62	+·1	+·2	·0	+·1	·0	·0	·0	−·1	·0	−·2	+·1	−·2
64	+·1	+·2	·0	+·1	·0	·0	·0	−·1	+·1	−·2	+·1	−·2
66	+·1	+·3	·0	+·2	·0	·0	·0	−·1	+·1	−·2	+·1	−·3

Month	a_2	b_2	a_2	b_2	a_2	b_2	a_2	b_2	a_2	b_2	a_2	b_2
Jan.	+·1	−·1	+·1	·0	+·1	·0	+·1	·0	+·1	+·1	+·1	+·1
Feb.	·0	−·2	+·1	−·2	+·1	−·2	+·2	−·1	+·2	−·1	+·2	·0
Mar.	−·1	−·3	·0	−·3	+·1	−·3	+·1	−·3	+·2	−·2	+·2	−·2
Apr.	−·2	−·3	−·2	−·3	−·1	−·4	·0	−·4	+·1	−·3	+·2	−·3
May	−·4	−·2	−·3	−·3	−·2	−·3	−·1	−·4	·0	−·4	+·1	−·4
June	−·4	·0	−·4	−·1	−·3	−·2	−·3	−·3	−·2	−·4	−·1	−·4
July	−·4	+·1	−·4	·0	−·4	−·1	−·3	−·2	−·3	−·3	−·2	−·3
Aug.	−·2	+·2	−·3	+·2	−·3	+·1	−·3	·0	−·3	−·1	−·3	−·2
Sept.	−·1	+·3	−·1	+·3	−·2	+·2	−·3	+·2	−·3	+·1	−·3	·0
Oct.	+·1	+·3	·0	+·3	·0	+·3	−·1	+·3	−·2	+·3	−·3	+·2
Nov.	+·3	+·2	+·2	+·3	+·1	+·4	·0	+·4	−·1	+·4	−·1	+·4
Dec.	+·4	+·1	+·4	+·2	+·3	+·3	+·2	+·4	+·1	+·4	·0	+·4

Latitude = Corrected observed altitude of *Polaris* + $a_0 + a_1 + a_2$

Azimuth of *Polaris* = $(b_0 + b_1 + b_2) / \cos(\text{latitude})$

LST	6^h		7^h		8^h		9^h		10^h		11^h	
	a_0	b_0	a_0	b_0	a_0	b_0	a_0	b_0	a_0	b_0	a_0	b_0
m	′	′	′	′	′	′	′	′	′	′	′	′
0	−26·9	−34·3	−17·1	−40·1	− 6·1	−43·1	+ 5·2	−43·2	+ 16·2	−40·3	+ 26·1	−34·7
3	26·5	34·7	16·6	40·3	5·6	43·2	5·8	43·1	16·7	40·0	26·5	34·3
6	26·0	35·0	16·0	40·5	5·0	43·3	6·3	43·0	17·2	39·8	27·0	34·0
9	25·5	35·4	15·5	40·8	4·4	43·3	6·9	42·9	17·8	39·6	27·4	33·6
12	25·1	35·7	15·0	40·9	3·9	43·4	7·5	42·8	18·3	39·4	27·8	33·3
15	−24·6	−36·0	−14·4	−41·1	− 3·3	−43·4	+ 8·0	−42·7	+ 18·8	−39·1	+ 28·3	−32·9
18	24·1	36·3	13·9	41·3	2·7	43·4	8·6	42·6	19·3	38·9	28·7	32·5
21	23·7	36·7	13·4	41·5	2·2	43·5	9·1	42·5	19·8	38·6	29·1	32·2
24	23·2	37·0	12·8	41·7	1·6	43·5	9·7	42·3	20·3	38·3	29·5	31·8
27	22·7	37·3	12·3	41·8	1·0	43·5	10·2	42·2	20·8	38·1	30·0	31·4
30	−22·2	−37·6	−11·7	−42·0	− 0·5	−43·5	+ 10·8	−42·1	+ 21·3	−37·8	+ 30·4	−31·0
33	21·7	37·8	11·2	42·1	+ 0·1	43·5	11·3	41·9	21·8	37·5	30·8	30·6
36	21·2	38·1	10·6	42·3	0·7	43·5	11·9	41·8	22·3	37·2	31·2	30·2
39	20·7	38·4	10·1	42·4	1·2	43·5	12·4	41·6	22·8	36·9	31·6	29·8
42	20·2	38·7	9·5	42·5	1·8	43·5	13·0	41·4	23·3	36·6	31·9	29·4
45	−19·7	−38·9	− 9·0	−42·6	+ 2·4	−43·4	+ 13·5	−41·3	+ 23·7	−36·3	+ 32·3	−28·9
48	19·2	39·2	8·4	42·7	2·9	43·4	14·1	41·1	24·2	36·0	32·7	28·5
51	18·7	39·4	7·8	42·9	3·5	43·3	14·6	40·9	24·7	35·7	33·1	28·1
54	18·1	39·7	7·3	42·9	4·1	43·3	15·1	40·7	25·1	35·3	33·4	27·7
57	17·6	39·9	6·7	43·0	4·6	43·2	15·7	40·5	25·6	35·0	33·8	27·2
60	−17·1	−40·1	− 6·1	−43·1	+ 5·2	−43·2	+ 16·2	−40·3	+ 26·1	−34·7	+ 34·1	−26·8

Lat.	a_1	b_1	a_1	b_1	a_1	b_1	a_1	b_1	a_1	b_1	a_1	b_1
°												
0	−·2	+·3	−·3	+·2	−·3	·0	−·3	−·2	−·3	−·3	−·2	−·3
10	−·2	+·2	−·3	+·1	−·3	·0	−·3	−·1	−·2	−·2	−·1	−·3
20	−·2	+·2	−·2	+·1	−·2	·0	−·2	−·1	−·2	−·2	−·1	−·2
30	−·1	+·1	−·2	+·1	−·2	·0	−·2	−·1	−·1	−·1	−·1	−·2
40	−·1	+·1	−·1	+·1	−·1	·0	−·1	·0	−·1	−·1	−·1	−·1
45	·0	·0	·0	·0	−·1	·0	·0	·0	·0	·0	·0	−·1
50	·0	·0	·0	·0	·0	·0	·0	·0	·0	·0	·0	·0
55	·0	−·1	+·1	·0	+·1	·0	+·1	·0	·0	+·1	·0	+·1
60	+·1	−·1	+·1	−·1	+·1	·0	+·1	+·1	+·1	+·1	+·1	+·1
62	+·1	−·2	+·2	−·1	+·2	·0	+·2	+·1	+·1	+·2	+·1	+·2
64	+·2	−·2	+·2	−·1	+·2	·0	+·2	+·1	+·2	+·2	+·1	+·2
66	+·2	−·3	+·3	−·2	+·3	·0	+·3	+·1	+·2	+·2	+·2	+·3

Month	a_2	b_2	a_2	b_2	a_2	b_2	a_2	b_2	a_2	b_2	a_2	b_2
Jan.	+·1	+·1	·0	+·1	·0	+·1	·0	+·1	−·1	+·1	−·1	+·1
Feb.	+·2	·0	+·2	+·1	+·2	+·1	+·1	+·2	+·1	+·2	·0	+·2
Mar.	+·3	−·1	+·3	·0	+·3	+·1	+·3	+·1	+·2	+·2	+·2	+·2
Apr.	+·3	−·2	+·3	−·2	+·4	−·1	+·4	·0	+·3	+·1	+·3	+·2
May	+·2	−·4	+·3	−·3	+·3	−·2	+·4	−·1	+·4	·0	+·4	+·1
June	·0	−·4	+·1	−·4	+·2	−·3	+·3	−·3	+·4	−·2	+·4	−·1
July	−·1	−·4	·0	−·4	+·1	−·4	+·2	−·3	+·3	−·3	+·3	−·2
Aug.	−·2	−·2	−·2	−·3	−·1	−·3	·0	−·3	+·1	−·3	+·2	−·3
Sept.	−·3	−·1	−·3	−·1	−·2	−·2	−·2	−·3	−·1	−·3	·0	−·3
Oct.	−·3	+·1	−·3	·0	−·3	·0	−·3	−·1	−·3	−·2	−·2	−·3
Nov.	−·2	+·3	−·3	+·2	−·4	+·1	−·4	·0	−·4	−·1	−·4	−·1
Dec.	−·1	+·4	−·2	+·4	−·3	+·3	−·4	+·2	−·4	+·1	−·4	·0

Latitude = Corrected observed altitude of *Polaris* + a_0 + a_1 + a_2

Azimuth of *Polaris* = (b_0 + b_1 + b_2) / cos (latitude)

LST	12^h a_0	b_0	13^h a_0	b_0	14^h a_0	b_0	15^h a_0	b_0	16^h a_0	b_0	17^h a_0	b_0
m	′	′	′	′	′	′	′	′	′	′	′	′
0	+34.1	−26.8	+39.9	−17.1	+43.0	−6.4	+43.2	+4.8	+40.5	+15.7	+35.1	+25.5
3	34.5	26.3	40.2	16.6	43.1	5.8	43.2	5.4	40.3	16.2	34.8	26.0
6	34.8	25.9	40.4	16.1	43.2	5.3	43.1	5.9	40.1	16.7	34.4	26.4
9	35.2	25.4	40.6	15.6	43.2	4.7	43.0	6.5	39.9	17.2	34.1	26.9
12	35.5	25.0	40.8	15.1	43.3	4.1	42.9	7.0	39.7	17.8	33.7	27.3
15	+35.8	−24.5	+41.0	−14.5	+43.3	−3.6	+42.8	+7.6	+39.4	+18.3	+33.4	+27.8
18	36.1	24.1	41.2	14.0	43.4	3.0	42.7	8.1	39.2	18.8	33.0	28.2
21	36.5	23.6	41.3	13.5	43.4	2.5	42.6	8.7	38.9	19.3	32.6	28.6
24	36.8	23.1	41.5	12.9	43.5	1.9	42.5	9.2	38.7	19.8	32.2	29.0
27	37.1	22.6	41.7	12.4	43.5	1.3	42.4	9.8	38.4	20.3	31.9	29.5
30	+37.4	−22.1	+41.8	−11.9	+43.5	−0.8	+42.2	+10.3	+38.1	+20.8	+31.5	+29.9
33	37.6	21.7	42.0	11.3	43.5	−0.2	42.1	10.9	37.9	21.3	31.1	30.3
36	37.9	21.2	42.1	10.8	43.5	+0.3	42.0	11.4	37.6	21.8	30.7	30.7
39	38.2	20.7	42.3	10.2	43.5	0.9	41.8	12.0	37.3	22.2	30.3	31.1
42	38.5	20.2	42.4	9.7	43.5	1.5	41.6	12.5	37.0	22.7	29.9	31.5
45	+38.7	−19.7	+42.5	−9.1	+43.5	+2.0	+41.5	+13.0	+36.7	+23.2	+29.5	+31.8
48	39.0	19.2	42.6	8.6	43.4	2.6	41.3	13.6	36.4	23.7	29.0	32.2
51	39.2	18.7	42.7	8.0	43.4	3.1	41.1	14.1	36.1	24.1	28.6	32.6
54	39.5	18.2	42.8	7.5	43.3	3.7	40.9	14.6	35.8	24.6	28.2	33.0
57	39.7	17.7	42.9	6.9	43.3	4.3	40.7	15.2	35.4	25.1	27.7	33.3
60	+39.9	−17.1	+43.0	−6.4	+43.2	+4.8	+40.5	+15.7	+35.1	+25.5	+27.3	+33.7

Lat.	a_1	b_1	a_1	b_1	a_1	b_1	a_1	b_1	a_1	b_1	a_1	b_1
°												
0	−.1	−.3	.0	−.2	.0	.0	.0	+.2	−.1	+.3	−.2	+.3
10	−.1	−.2	.0	−.1	.0	.0	.0	+.1	−.1	+.2	−.1	+.3
20	−.1	−.2	.0	−.1	.0	.0	.0	+.1	−.1	+.2	−.1	+.2
30	.0	−.1	.0	−.1	.0	.0	.0	+.1	.0	+.1	−.1	+.2
40	.0	−.1	.0	−.1	.0	.0	.0	.0	.0	+.1	.0	+.1
45	.0	.0	.0	.0	.0	.0	.0	.0	.0	.0	.0	+.1
50	.0	.0	.0	.0	.0	.0	.0	.0	.0	.0	.0	.0
55	.0	+.1	.0	.0	.0	.0	.0	.0	.0	−.1	.0	−.1
60	.0	+.1	.0	+.1	.0	.0	.0	−.1	.0	−.1	+.1	−.1
62	+.1	+.2	.0	+.1	.0	.0	.0	−.1	.0	−.2	+.1	−.2
64	+.1	+.2	.0	+.1	.0	.0	.0	−.1	+.1	−.2	+.1	−.2
66	+.1	+.3	.0	+.2	.0	.0	.0	−.1	+.1	−.2	+.1	−.3

Month	a_2	b_2	a_2	b_2	a_2	b_2	a_2	b_2	a_2	b_2	a_2	b_2
Jan.	−.1	+.1	−.1	.0	−.1	.0	−.1	.0	−.1	−.1	−.1	−.1
Feb.	.0	+.2	−.1	+.2	−.1	+.2	−.2	+.1	−.2	+.1	−.2	.0
Mar.	+.1	+.3	.0	+.3	−.1	+.3	−.1	+.3	−.2	+.2	−.2	+.2
Apr.	+.2	+.3	+.2	+.3	+.1	+.4	.0	+.4	−.1	+.3	−.2	+.3
May	+.4	+.2	+.3	+.3	+.2	+.3	+.1	+.4	.0	+.4	−.1	+.4
June	+.4	.0	+.4	+.1	+.3	+.2	+.3	+.3	+.2	+.4	+.1	+.4
July	+.4	−.1	+.4	.0	+.4	+.1	+.3	+.2	+.3	+.3	+.2	+.3
Aug.	+.2	−.2	+.3	−.2	+.3	−.1	+.3	.0	+.3	+.1	+.3	+.2
Sept.	+.1	−.3	+.1	−.3	+.2	−.2	+.3	−.2	+.3	−.1	+.3	.0
Oct.	−.1	−.3	.0	−.3	.0	−.3	+.1	−.3	+.2	−.3	+.3	−.2
Nov.	−.3	−.2	−.2	−.3	−.1	−.4	.0	−.4	+.1	−.4	+.1	−.4
Dec.	−.4	−.1	−.4	−.2	−.3	−.3	−.2	−.4	−.1	−.4	.0	−.4

Latitude = Corrected observed altitude of *Polaris* + a_0 + a_1 + a_2

Azimuth of *Polaris* = $(b_0 + b_1 + b_2) / \cos(\text{latitude})$

LST	18^h		19^h		20^h		21^h		22^h		23^h	
	a_0	b_0	a_0	b_0	a_0	b_0	a_0	b_0	a_0	b_0	a_0	b_0
m	′	′	′	′	′	′	′	′	′	′	′	′
0	+ 27·3	+ 33·7	+ 17·7	+ 39·6	+ 6·8	+ 42·9	− 4·6	+ 43·3	− 15·6	+ 40·7	− 25·6	+ 35·3
3	26·9	34·1	17·1	39·9	6·2	43·0	5·1	43·2	16·2	40·5	26·1	35·0
6	26·4	34·4	16·6	40·1	5·7	43·1	5·7	43·2	16·7	40·3	26·5	34·6
9	26·0	34·7	16·1	40·3	5·1	43·2	6·3	43·1	17·2	40·1	27·0	34·3
12	25·5	35·1	15·6	40·5	4·5	43·2	6·8	43·0	17·7	39·8	27·4	33·9
15	+ 25·0	+ 35·4	+ 15·0	+ 40·7	+ 4·0	+ 43·3	− 7·4	+ 42·9	− 18·3	+ 39·6	− 27·9	+ 33·5
18	24·6	35·7	14·5	40·9	3·4	43·3	7·9	42·8	18·8	39·4	28·3	33·2
21	24·1	36·1	14·0	41·1	2·8	43·4	8·5	42·7	19·3	39·1	28·8	32·8
24	23·6	36·4	13·4	41·3	2·3	43·4	9·1	42·6	19·8	38·9	29·2	32·4
27	23·2	36·7	12·9	41·5	1·7	43·5	9·6	42·5	20·3	38·6	29·6	32·0
30	+ 22·7	+ 37·0	+ 12·3	+ 41·6	+ 1·1	+ 43·5	− 10·2	+ 42·4	− 20·8	+ 38·3	− 30·0	+ 31·6
33	22·2	37·3	11·8	41·8	+ 0·6	43·5	10·7	42·2	21·3	38·1	30·4	31·2
36	21·7	37·6	11·2	41·9	0·0	43·5	11·3	42·1	21·8	37·8	30·8	30·8
39	21·2	37·8	10·7	42·1	− 0·6	43·5	11·8	41·9	22·3	37·5	31·2	30·4
42	20·7	38·1	10·1	42·2	1·2	43·5	12·4	41·8	22·8	37·2	31·6	30·0
45	+ 20·2	+ 38·4	+ 9·6	+ 42·4	− 1·7	+ 43·5	− 12·9	+ 41·6	− 23·3	+ 36·9	− 32·0	+ 29·6
48	19·7	38·7	9·0	42·5	2·3	43·5	13·5	41·5	23·7	36·6	32·4	29·2
51	19·2	38·9	8·5	42·6	2·9	43·4	14·0	41·3	24·2	36·3	32·8	28·7
54	18·7	39·2	7·9	42·7	3·4	43·4	14·6	41·1	24·7	36·0	33·2	28·3
57	18·2	39·4	7·3	42·8	4·0	43·3	15·1	40·9	25·2	35·6	33·5	27·9
60	+ 17·7	+ 39·6	+ 6·8	+ 42·9	− 4·6	+ 43·3	− 15·6	+ 40·7	− 25·6	+ 35·3	− 33·9	+ 27·4
Lat.	a_1	b_1	a_1	b_1	a_1	b_1	a_1	b_1	a_1	b_1	a_1	b_1
°												
0	− ·2	+ ·3	− ·3	+ ·2	− ·3	·0	− ·3	− ·2	− ·3	− ·3	− ·2	− ·3
10	− ·2	+ ·2	− ·3	+ ·1	− ·3	·0	− ·3	− ·1	− ·2	− ·2	− ·1	− ·3
20	− ·2	+ ·2	− ·2	+ ·1	− ·2	·0	− ·2	− ·1	− ·2	− ·2	− ·1	− ·2
30	− ·1	+ ·1	− ·2	+ ·1	− ·2	·0	− ·2	− ·1	− ·1	− ·1	− ·1	− ·2
40	− ·1	+ ·1	− ·1	+ ·1	− ·1	·0	− ·1	·0	− ·1	− ·1	− ·1	− ·1
45	·0	·0	·0	·0	− ·1	·0	·0	·0	·0	·0	·0	− ·1
50	·0	·0	·0	·0	·0	·0	·0	·0	·0	·0	·0	·0
55	·0	− ·1	+ ·1	·0	+ ·1	·0	+ ·1	·0	·0	+ ·1	·0	+ ·1
60	+ ·1	− ·1	+ ·1	− ·1	+ ·1	·0	+ ·1	+ ·1	+ ·1	+ ·1	+ ·1	+ ·1
62	+ ·1	− ·2	+ ·2	− ·1	+ ·2	·0	+ ·2	+ ·1	+ ·1	+ ·2	+ ·1	+ ·2
64	+ ·2	− ·2	+ ·2	− ·1	+ ·2	·0	+ ·2	+ ·1	+ ·2	+ ·2	+ ·1	+ ·2
66	+ ·2	− ·3	+ ·3	− ·2	+ ·3	·0	+ ·3	+ ·1	+ ·2	+ ·2	+ ·2	+ ·3
Month	a_2	b_2	a_2	b_2	a_2	b_2	a_2	b_2	a_2	b_2	a_2	b_2
Jan.	− ·1	− ·1	·0	− ·1	·0	− ·1	·0	− ·1	+ ·1	− ·1	+ ·1	− ·1
Feb.	− ·2	·0	− ·2	− ·1	− ·2	− ·1	− ·1	− ·2	− ·1	− ·2	·0	− ·2
Mar.	− ·3	+ ·1	− ·3	·0	− ·3	− ·1	− ·3	− ·1	− ·2	− ·2	− ·2	− ·2
Apr.	− ·3	+ ·2	− ·3	+ ·2	− ·4	+ ·1	− ·4	·0	− ·3	− ·1	− ·3	− ·2
May	− ·2	+ ·4	− ·3	+ ·3	− ·3	+ ·2	− ·4	+ ·1	− ·4	·0	− ·4	− ·1
June	·0	+ ·4	− ·1	+ ·4	− ·2	+ ·3	− ·3	+ ·3	− ·4	+ ·2	− ·4	+ ·1
July	+ ·1	+ ·4	·0	+ ·4	− ·1	+ ·4	− ·2	+ ·3	− ·3	+ ·3	− ·3	+ ·2
Aug.	+ ·2	+ ·2	+ ·2	+ ·3	+ ·1	+ ·3	·0	+ ·3	− ·1	+ ·3	− ·2	+ ·3
Sept.	+ ·3	+ ·1	+ ·3	+ ·1	+ ·2	+ ·2	+ ·2	+ ·3	+ ·1	+ ·3	·0	+ ·3
Oct.	+ ·3	− ·1	+ ·3	·0	+ ·3	·0	+ ·3	+ ·1	+ ·3	+ ·2	+ ·2	+ ·3
Nov.	+ ·2	− ·3	+ ·3	− ·2	+ ·4	− ·1	+ ·4	·0	+ ·4	+ ·1	+ ·4	+ ·1
Dec.	+ ·1	− ·4	+ ·2	− ·4	+ ·3	− ·3	+ ·4	− ·2	+ ·4	− ·1	+ ·4	·0

Latitude = Corrected observed altitude of *Polaris* + a_0 + a_1 + a_2

Azimuth of *Polaris* = $(b_0 + b_1 + b_2) / \cos(\text{latitude})$

Call sign Station	Frequency (kHz)	Power (kW)	Notes on signals (All times are in UTC)
MSF Rugby England	* 60	27	Continuous except 10^h–14^h on first Tuesday of each month, 1^h later in Summer (DST). Second markers 100 ms; minute markers 500 ms; both markers interruption of carrier. Time codes for automatic equipment (a) by lengthening or doubling of some second markers from 17^s to 59^s and (b) within the minute marker.
DCF77 Mainflingen Germany	77·5	38	Continuous. Second markers of 100 ms and 200 ms; markers interruption of carrier; no interruption at 59^s.
ZSC Cape Town South Africa	418 4 291 8 461 12 772·5 17 018	5 10 10 10 10	07^h 55^m–08^h, 16^h 55^m–17^h. Second markers 100 ms, minute markers 400 ms.
RWM Moscow Russia	* 4 996 9 996 14 996	5 5 8	Continuous except between 05^h–13^h on some Wednesdays. Second markers 40 ms and 100 ms, minute markers 500 ms. Time signals between 10^m–30^m and 40^m–60^m. DUT1 coded between 10^m–20^m and 40^m–50^m.
JJY Sanwa Japan	* 2 500 5 000 8 000 10 000 15 000	2 2 2 2 2	Continuous. Second markers 5 ms of 1·6 kHz tone, minute *warning* markers 655 ms. Silence between 35^m–39^m. Voice and morse announcement of time (UT + 9^h) every 10^m.
VNG Llandilo Penrith Australia	* 2 500 5 000 8 638 12 984 16 000	1 10 10 10 5	Continuous except for 16 000 kHz frequency, which transmits 22^h–10^h. Markers are tone pulses of 1 kHz; minute markers 500 ms, second markers normally 50 ms; length coded for date and time between 21^s–46^s; second marker at 59^s omitted.
WWVH Kauai, Kekaha Hawaii U.S.A.	* 2 500 5 000 10 000 15 000	5 10 10 10	Continuous. Second markers 5 ms of 1·2 kHz tone, minute markers 800 ms of 1·2 kHz tone; hour markers 800 ms of 1·5 kHz tone; no audio tone between 08^m–12^m, 14^m–20^m. Female voice identification at 00^m and 30^m.
WWV Fort Collins Colorado U.S.A.	* 2 500 5 000 10 000 15 000 20 000	2·5 10 10 10 2·5	Continuous. Second markers 5 ms of 1 kHz tone, minute markers 800 ms of 1 kHz tone, hour markers 800 ms of 1·5 kHz tone; no audio tone between 43^m–52^m. Male voice identification at 00^m and 30^m.
CHU Ottawa Canada	* 3 330 7 335 14 670	3 10 3	Continuous. Second markers 300 ms, minute markers 500 ms, hour markers 1^s; silence at 29^s. Voice announcement of time (UT -5^h) in French and English between 51^s–59^s.
LOL Buenos Aires Argentina	* 5 000 10 000 15 000	2 2 2	11^h–12^h, 14^h–15^h, 17^h–18^h, 20^h–21^h, 23^h–24^h. Second markers 5 ms with either 1 kHz or 440 Hz tone; 59^s marker omitted. Voice announcement of time (UT-3^h).

* See note opposite.

Call sign Station	Frequency (kHz)	Power (kW)	Notes on signals (All times are in UTC)
LOL	* 4 856		00^h 55^m–01^h, 12^h 55^m–13^h, 20^h 55^m–21^h.
Buenos Aires	8 030		Markers as for frequencies on previous page, except that
Argentina	17 180		the 29^s marker is omitted. English voice announcement.

NOTES ON RADIO TIME SIGNALS

The list of radio time signals is restricted to the principal signals that are likely to be used by land surveyors. The details of the transmissions are believed to be correct at February 2001, but the signals are liable to changes at short notice. An extended list of radio time signals is given in *The Admiralty List of Radio Signals*, Volume 2; this is published by the Hydrographer of the Navy and is obtainable from the Agents for the sale of Admiralty Charts. Details of amendments to the signals are given in Section VI of the Weekly Edition of *Admiralty Notices to Mariners* (published as for the Admiralty List).

The signals in the list are all based on the system of coordinated universal time (UTC) that was introduced in 1972 January 1. This system is derived from the international atomic time scale (TAI), and differs from it only by an integral number of seconds, but step adjustments of 1 second are made from time to time in order that UTC shall not differ from UT1, (simply denoted by UT elsewhere in this Almanac), which corresponds to mean solar time on the meridian of Greenwich, by more than 0·9 seconds. Any such step adjustments are normally made at the last UTC second of December 31 and June 30, but may be made at the end of any other month if this is necessary in order to maintain compatibility with UT1. The signals indicated (*) are coded by emphasising some seconds markers to give DUT1, the predicted value of UT1 − UTC rounded to a precision of 0·1 seconds. If the seconds markers from 1^s to n^s are emphasised, the difference is $+ n \times$ 0·1 seconds; if the seconds markers from 9^s to $(8 + m)^s$ are emphasised, the difference is $- m \times$ 0·1 seconds (n and m are always less than or equal to 8). The appropriate seconds markers may be emphasised by lengthening, doubling, splitting or tone modulation of the normal seconds markers. In reducing observations it is possible either to apply corrections to all the times to reduce them to UT1 (e.g. by incorporating them with the corrections to the chronometer) or to apply corrections to the deduced positions. The correction in longitude in seconds of arc, measured positively to the east, is equal to *minus* fifteen times the difference UT1 − UTC in seconds of time. Provisional values in seconds (s) of the differences UT1 − UTC and TAI − UTC, are given in the weekly *International Earth Rotation Service (IERS) Bulletin A*, obtainable from National Earth Orientation Service, US Naval Observatory, Washington DC 20392 − 5100, United States of America. Their World Wide Web address on the Internet is http://maia.usno.navy.mil/.

In all cases the commencement of each minute and second marker indicates the timing reference point. In the case of JJY a lengthened additional marker gives previous notice that the next second is the minute marker. An A1 type signal consists of a continuous, unmodulated, carrier wave with on-off keying.

Warning: The "6-pips" time-signals broadcast in the World Service of the BBC are not intended for precise use. In direct transmissions from the UK the start of the long final pip will normally be received within 0^{s}1 of UTC, but signals from relay stations abroad may be subject to additional delays of 0^{s}25.

REFRACTION TABLES

MEAN REFRACTION, r_0

(Critical table: the r_0 value given applies between the two bracketing altitudes. Alt. in ° ′, r_0 in ″.)

Alt.	r_0
10 00	319
10 01	318
10 03	317
10 05	316
10 07	315
10 09	314
10 11	313
10 13	312
10 15	311
10 17	310
10 19	309
10 21	308
10 23	307
10 25	306
10 27	305
10 29	304
10 32	303
10 34	302
10 36	301
10 38	300
10 40	299
10 43	298
10 45	297
10 47	296
10 49	295
10 52	294
10 54	293
10 56	292
10 58	291
11 01	290
11 03	289
11 06	288
11 08	287
11 10	286
11 13	285
11 15	284
11 18	283
11 20	282
11 23	281
11 25	280
11 28	279
11 30	278
11 33	277
11 35	276
11 38	275
11 40	274
11 43	273
11 46	

Alt.	r_0
11 43	273
11 46	272
11 48	271
11 51	270
11 54	269
11 56	268
11 59	267
12 02	266
12 05	265
12 07	264
12 10	263
12 13	262
12 16	261
12 19	260
12 22	259
12 25	258
12 27	257
12 30	256
12 33	255
12 36	254
12 39	253
12 42	252
12 45	251
12 49	250
12 52	249
12 55	248
12 58	247
13 01	246
13 04	245
13 08	244
13 11	243
13 14	242
13 17	241
13 21	240
13 24	239
13 27	238
13 31	237
13 34	236
13 38	235
13 41	234
13 45	233
13 48	232
13 52	231
13 55	230
13 59	229
14 03	228
14 06	227
14 10	

Alt.	r_0
14 06	227
14 10	226
14 14	225
14 18	224
14 21	223
14 25	222
14 29	221
14 33	220
14 37	219
14 41	218
14 45	217
14 49	216
14 53	215
14 57	214
15 01	213
15 05	212
15 10	211
15 14	210
15 18	209
15 22	208
15 27	207
15 31	206
15 36	205
15 40	204
15 45	203
15 49	202
15 54	201
15 58	200
16 03	199
16 08	198
16 13	197
16 17	196
16 22	195
16 27	194
16 32	193
16 37	192
16 42	191
16 47	190
16 52	189
16 58	188
17 03	187
17 08	186
17 13	185
17 19	184
17 24	183
17 30	182
17 35	181
17 41	

Alt.	r_0
17 35	181
17 41	180
17 47	179
17 52	178
17 58	177
18 04	176
18 10	175
18 16	174
18 22	173
18 28	172
18 34	171
18 40	170
18 47	169
18 53	168
18 59	167
19 06	166
19 12	165
19 19	164
19 26	163
19 32	162
19 39	161
19 46	160
19 53	159
20 00	158
20 07	157
20 14	156
20 22	155
20 29	154
20 36	153
20 44	152
20 52	151
20 59	150
21 07	149
21 15	148
21 23	147
21 31	146
21 39	145
21 48	144
21 56	143
22 04	142
22 13	141
22 22	140
22 30	139
22 39	138
22 48	137
22 57	136
23 07	135
23 16	

Alt.	r_0
23 07	135
23 16	134
23 25	133
23 35	132
23 45	131
23 55	130
24 04	129
24 15	128
24 25	127
24 35	126
24 45	125
24 56	124
25 07	123
25 18	122
25 29	121
25 40	120
25 51	119
26 03	118
26 14	117
26 26	116
26 38	115
26 50	114
27 03	113
27 15	112
27 28	111
27 40	110
27 53	109
28 07	108
28 20	107
28 34	106
28 47	105
29 01	104
29 15	103
29 30	102
29 44	101
29 59	100
30 14	99
30 29	98
30 45	97
31 01	96
31 17	95
31 33	94
31 49	93
32 06	92
32 23	91
32 40	90
32 58	

Alt.	r_0
32 40	90
32 58	89
33 15	88
33 33	87
33 52	86
34 10	85
34 29	84
34 49	83
35 08	82
35 28	81
35 48	80
36 09	79
36 30	78
36 51	77
37 12	76
37 34	75
37 56	74
38 19	73
38 42	72
39 06	71
39 29	70
39 54	69
40 18	68
40 43	67
41 09	66
41 35	65
42 01	64
42 28	63
42 55	62
43 23	61
43 51	60
44 20	59
44 49	58
45 19	57
45 49	56
46 20	55
46 51	54
47 23	53
47 55	52
48 28	51
49 01	50
49 35	49
50 10	48
50 45	47
51 21	46
51 58	45
52 35	

Alt.	r_0
51 58	45
52 35	44
53 12	43
53 50	42
54 29	41
55 09	40
55 49	39
56 30	38
57 11	37
57 53	36
58 36	35
59 20	34
60 04	33
60 48	32
61 34	31
62 20	30
63 07	29
63 54	28
64 42	27
65 31	26
66 20	25
67 10	24
68 00	23
68 51	22
69 43	21
70 35	20
71 28	19
72 22	18
73 15	17
74 10	16
75 05	15
76 00	14
76 56	13
77 52	12
78 49	11
79 46	10
80 43	9
81 41	8
82 39	7
83 37	6
84 36	5
85 34	4
86 33	3
87 32	2
88 31	1
89 30	0
90 00	

In critical cases ascend.

$$\text{Refraction} = r_0 \times f$$

where r_0 is the mean refraction and f is the correcting factor on pages 63, 64.

CORRECTING FACTOR, ƒ, FOR PRESSURE AND TEMPERATURE

1080 °C	1070 °C	1060 °C	1050 °C	mb ƒ
		+58	+55	0·90
	+57	54	51	0·91
+57	53	50	47	0·92
53	50	47	44	0·93
49	46	43	40	0·94
46	43	40	37	0·95
43	40	37	34	0·96
39	36	34	31	0·97
36	33	30	28	0·98
33	30	27	25	0·99
30	27	24	22	1·00
27	24	21	19	1·01
24	21	19	16	1·02
21	18	16	13	1·03
18	16	13	10	1·04
16	13	10	8	1·05
13	10	8	5	1·06
10	8	5	+2	1·07
8	5	+3	0	1·08
5	+3	0	−3	1·09
+3	0	−2	5	1·10
0	−2	5	7	1·11
−2	5	7	10	1·12
5	7	10	12	1·13
7	9	12	14	1·14
9	12	14	17	1·15
12	14	16	19	1·16
−14	−16	−19	−21	

1040 °C	1030 °C	1020 °C	1010 °C	mb ƒ
		+56	+53	0·87
+59	+56	52	49	0·88
55	52	49	46	0·89
51	48	45	42	0·90
48	45	42	38	0·91
44	41	38	35	0·92
41	38	35	32	0·93
37	34	31	29	0·94
34	31	28	25	0·95
31	28	25	22	0·96
28	25	22	19	0·97
25	22	19	16	0·98
22	19	16	13	0·99
19	16	13	11	1·00
16	13	10	8	1·01
13	10	8	5	1·02
10	8	5	+2	1·03
8	5	+2	0	1·04
5	+2	0	−3	1·05
+2	0	−3	5	1·06
0	−3	5	8	1·07
−3	5	8	10	1·08
5	8	10	13	1·09
8	10	13	15	1·10
10	12	15	17	1·11
12	15	17	20	1·12
15	17	19	22	1·13
−17	−19	−22	−24	

1000 °C	990 °C	980 °C	970 °C	mb ƒ
+54	+50	+47	+44	0·86
50	46	43	40	0·87
46	43	40	36	0·88
42	39	36	33	0·89
39	36	33	29	0·90
35	32	29	26	0·91
32	29	26	23	0·92
29	26	23	20	0·93
26	23	20	17	0·94
22	19	17	14	0·95
19	16	14	11	0·96
16	13	11	8	0·97
13	11	8	5	0·98
11	8	5	+2	0·99
8	5	+2	−1	1·00
5	+2	−1	3	1·01
+2	0	3	6	1·02
0	−3	6	8	1·03
−3	6	8	11	1·04
6	8	11	13	1·05
8	11	13	16	1·06
10	13	16	18	1·07
13	16	18	21	1·08
15	18	20	23	1·09
18	20	23	−25	1·10
20	22	−25		1·11
22	−25			1·12
−24				

960 / 450 °C	950 / 550 °C	940 / 650 °C	930 / 700 °C	mb m ƒ
+48	+45	+41	+38	0·84
44	41	37	34	0·85
40	37	34	31	0·86
37	34	30	27	0·87
33	30	27	24	0·88
30	27	24	20	0·89
26	23	20	17	0·90
23	20	17	14	0·91
20	17	14	11	0·92
17	14	11	8	0·93
14	11	8	5	0·94
11	8	5	+2	0·95
8	5	+2	−1	0·96
5	+2	−1	4	0·97
+2	−1	4	6	0·98
−1	3	6	9	0·99
3	6	9	12	1·00
6	9	12	14	1·01
9	11	14	17	1·02
11	14	17	19	1·03
14	16	19	22	1·04
16	19	21	24	1·05
19	21	24	−26	1·06
21	24	−26		1·07
−23	−26			

920 / 800 °C	910 / 900 °C	900 / 1000 °C	890 / 1100 °C	mb m ƒ
+46	+43	+39	+36	0·81
42	39	35	32	0·82
38	35	32	28	0·83
35	31	28	25	0·84
31	28	24	21	0·85
27	24	21	18	0·86
24	21	18	14	0·87
21	17	14	11	0·88
17	14	11	8	0·89
14	11	8	5	0·90
11	8	5	+2	0·91
8	5	+2	−1	0·92
5	+2	−1	4	0·93
+2	−1	4	7	0·94
−1	4	7	10	0·95
4	7	9	12	0·96
6	9	12	15	0·97
9	12	15	18	0·98
12	15	17	20	0·99
14	17	20	23	1·00
17	20	23	25	1·01
20	22	25	28	1·02
22	25	27	−30	1·03
24	27	−30		1·04
−27	−29			

880 / 1150 °C	870 / 1250 °C	860 / 1350 °C	850 / 1450 °C	mb m ƒ
		+49	+46	0·75
	+49	45	42	0·76
+48	45	41	37	0·77
44	41	37	33	0·78
40	37	33	30	0·79
36	33	29	26	0·80
32	29	25	22	0·81
29	25	22	19	0·82
25	22	18	15	0·83
21	18	15	12	0·84
18	15	11	8	0·85
15	11	8	5	0·86
11	8	5	+2	0·87
8	5	+2	−1	0·88
5	+2	−1	4	0·89
+2	−1	4	7	0·90
−1	4	7	10	0·91
4	7	10	13	0·92
7	10	13	16	0·93
10	13	16	19	0·94
13	15	18	21	0·95
15	18	21	24	0·96
18	21	24	26	0·97
20	23	26	−29	0·98
−23	−26	−29		

REFRACTION TABLES

CORRECTING FACTOR, *f*, FOR PRESSURE AND TEMPERATURE

840	830	820	810	mb		800	790	780	770	mb		760	750	740	730	mb
1550	1650	1750	1850	m		1950	2050	2150	2250	m		2350	2450	2550	2700	m
°C	°C	°C	°C	*f*		°C	°C	°C	°C	*f*		°C	°C	°C	°C	*f*
	+52	+48	+44	0·72				+51	+47	0·68					+56	0·63
+51	47	43	40	0·73			+50	46	42	0·69				+55	50	0·64
46	43	39	35	0·74		+49	45	41	38	0·70			+54	49	45	0·65
42	38	35	31	0·75		45	41	37	33	0·71		+53	49	45	41	0·66
38	34	31	27	0·76		40	36	33	29	0·72		48	44	40	36	0·67
34	30	27	23	0·77		36	32	29	25	0·73		43	39	35	31	0·68
30	26	23	20	0·78		32	28	24	21	0·74		38	34	31	27	0·69
26	23	19	16	0·79		28	24	20	17	0·75		34	30	26	23	0·70
22	19	16	12	0·80		24	20	17	13	0·76		30	26	22	18	0·71
19	15	12	9	0·81		20	16	13	9	0·77		25	22	18	14	0·72
15	12	9	5	0·82		16	13	9	6	0·78		21	18	14	10	0·73
12	8	5	+2	0·83		12	9	6	+2	0·79		17	14	10	7	0·74
8	5	+2	−1	0·84		9	6	+2	−1	0·80		13	10	6	+3	0·75
5	+2	−1	4	0·85		5	+2	−1	4	0·81		10	6	+3	−1	0·76
+2	−1	4	8	0·86		+2	−1	4	8	0·82		6	+3	−1	4	0·77
−1	4	7	11	0·87		−1	4	8	11	0·83		+2	−1	4	8	0·78
4	7	10	14	0·88		4	8	11	14	0·84		−1	4	8	11	0·79
7	10	13	16	0·89		8	11	14	17	0·85		4	8	11	14	0·80
10	13	16	19	0·90		11	14	17	20	0·86		8	11	14	17	0·81
13	16	19	22	0·91		14	17	20	23	0·87		11	14	17	21	0·82
16	19	22	25	0·92		17	20	23	26	0·88		14	17	20	24	0·83
19	22	25	28	0·93		19	23	26	29	0·89		17	20	23	27	0·84
21	24	27	−30	0·94		22	25	28	−31	0·90		20	23	26	−29	0·85
24	27	−30		0·95		25	28	−31		0·91		23	26	−29		0·86
27	−30			0·96		28	−31			0·92		26	−29			0·87
−29						−30						−29				

720	710	700	690	mb		680	670	660	650	mb		640	630	620	610	mb
2800	2900	3000	3150	m		3250	3350	3450	3600	m		3700	3850	3950	4100	m
°C	°C	°C	°C	*f*		°C	°C	°C	°C	*f*		°C	°C	°C	°C	*f*
		+54	+49	0·61				+53	+49	0·58					+54	0·54
	+53	48	44	0·62			+52	47	43	0·59				+53	48	0·55
+51	47	43	39	0·63		+51	46	42	38	0·60			+51	47	42	0·56
46	42	38	34	0·64		45	41	37	33	0·61		+50	46	41	37	0·57
41	37	33	29	0·65		40	36	32	28	0·62		44	40	36	31	0·58
37	33	29	25	0·66		35	31	27	23	0·63		39	35	30	26	0·59
32	28	24	20	0·67		30	26	22	18	0·64		34	29	25	21	0·60
27	24	20	16	0·68		25	22	18	14	0·65		29	25	20	16	0·61
23	19	16	12	0·69		21	17	13	10	0·66		24	20	16	12	0·62
19	15	12	8	0·70		17	13	9	5	0·67		19	15	11	7	0·63
15	11	8	+4	0·71		12	9	5	+1	0·68		14	11	7	+3	0·64
11	7	+4	0	0·72		8	5	+1	−3	0·69		10	6	+2	−1	0·65
7	+3	0	−4	0·73		+4	+1	−3	7	0·70		6	+2	−2	5	0·66
+3	0	−4	7	0·74		0	−3	7	10	0·71		+2	−2	6	9	0·67
0	−4	7	11	0·75		−3	7	10	14	0·72		−2	6	10	13	0·68
−4	7	11	14	0·76		7	11	14	17	0·73		6	10	13	17	0·69
8	11	14	18	0·77		11	14	18	21	0·74		10	14	17	21	0·70
11	14	18	21	0·78		14	18	21	24	0·75		14	17	21	24	0·71
14	18	21	24	0·79		18	21	24	28	0·76		17	21	24	28	0·72
17	21	24	27	0·80		21	24	28	−31	0·77		21	24	28	−31	0·73
21	24	27	−30	0·81		24	27	−31		0·78		24	28	−31		0·74
24	27	−30		0·82		27	−31			0·79		28	−31			0·75
−27	−30					−30						−31				

In critical cases ascend.

° ′		° ′		° ′		° ′		° ′		° ′	
0 00·0	1·000	12 48·2	1·027	18 10·3	1·054	22 07·5	1·081	25 20·6	1·108	28 05·2	1·135
3 08·1	·003	13 31·1	·030	18 39·7	·057	22 30·8	·084	25 40·1	·111	28 22·2	·138
5 25·5	·006	14 11·7	·033	19 08·2	·060	22 53·5	·087	25 59·3	·114	28 38·9	·141
6 59·7	·009	14 50·3	·036	19 35·9	·063	23 15·8	·090	26 18·2	·117	28 55·3	·144
8 16·0	·012	15 27·1	·039	20 02·8	·066	23 37·6	·093	26 36·7	·120	29 11·5	·147
9 21·7	·015	16 02·3	·042	20 29·0	·069	23 59·0	·096	26 55·0	·123	29 27·5	·150
10 20·2	·018	16 36·1	·045	20 54·5	·072	24 20·0	·099	27 12·9	·126	29 43·3	·153
11 13·4	·021	17 08·6	·048	21 19·5	·075	24 40·6	·102	27 30·6	·129	29 58·9	·156
12 02·5	1·024	17 40·0	·051	21 43·8	1·078	25 00·8	1·105	27 48·1	1·132	30 14·3	1·159
12 48·2		18 10·3	1·051	22 07·5		25 20·6		28 05·2		30 29·5	

In critical cases ascend.

°	00′	05′	10′	15′	20′	25′	30′	35′	40′	45′	50′	55′
30	1·155	1·156	1·157	1·158	1·159	1·160	1·161	1·162	1·163	1·164	1·165	1·166
31	·167	·168	·169	·170	·171	·172	·173	·174	·175	·176	·177	·178
32	·179	·180	·181	·182	·184	·185	·186	·187	·188	·189	·190	·191
33	·192	·193	·195	·196	·197	·198	·199	·200	·202	·203	·204	·205
34	·206	·207	·209	·210	·211	·212	·213	·215	·216	·217	·218	·220
35	1·221	1·222	1·223	1·225	1·226	1·227	1·228	1·230	1·231	1·232	1·233	1·235
36	·236	·237	·239	·240	·241	·243	·244	·245	·247	·248	·249	·251
37	·252	·254	·255	·256	·258	·259	·260	·262	·263	·265	·266	·268
38	·269	·270	·272	·273	·275	·276	·278	·279	·281	·282	·284	·285
39	·287	·288	·290	·291	·293	·294	·296	·298	·299	·301	·302	·304
40	1·305	1·307	1·309	1·310	1·312	1·313	1·315	1·317	1·318	1·320	1·322	1·323
41	·325	·327	·328	·330	·332	·333	·335	·337	·339	·340	·342	·344
42	·346	·347	·349	·351	·353	·355	·356	·358	·360	·362	·364	·365
43	·367	·369	·371	·373	·375	·377	·379	·381	·382	·384	·386	·388
44	·390	·392	·394	·396	·398	·400	·402	·404	·406	·408	·410	·412
45	1·414	1·416	1·418	1·420	1·423	1·425	1·427	1·429	1·431	1·433	1·435	1·437
46	·440	·442	·444	·446	·448	·451	·453	·455	·457	·459	·462	·464
47	·466	·469	·471	·473	·476	·478	·480	·483	·485	·487	·490	·492
48	·494	·497	·499	·502	·504	·507	·509	·512	·514	·517	·519	·522
49	·524	·527	·529	·532	·535	·537	·540	·542	·545	·548	·550	·553
50	1·556	1·558	1·561	1·564	1·567	1·569	1·572	1·575	1·578	1·581	1·583	1·586
51	·589	·592	·595	·598	·601	·603	·606	·609	·612	·615	·618	·621
52	·624	·627	·630	·633	·636	·640	·643	·646	·649	·652	·655	·658
53	·662	·665	·668	·671	·675	·678	·681	·684	·688	·691	·695	·698
54	·701	·705	·708	·712	·715	·719	·722	·726	·729	·733	·736	·740
55	1·743	1·747	1·751	1·754	1·758	1·762	1·766	1·769	1·773	1·777	1·781	1·784
56	·788	·792	·796	·800	·804	·808	·812	·816	·820	·824	·828	·832
57	·836	·840	·844	·849	·853	·857	·861	·865	·870	·874	·878	·883
58	·887	·891	·896	·900	·905	·909	·914	·918	·923	·928	·932	·937
59	1·942	1·946	1·951	1·956	1·961	1·965	1·970	1·975	1·980	1·985	1·990	1·995
60	2·000	2·005	2·010	2·015	2·020	2·026	2·031	2·036	2·041	2·047	2·052	2·057
61	·063	·068	·074	·079	·085	·090	·096	·101	·107	·113	·118	·124
62	·130	·136	·142	·148	·154	·160	·166	·172	·178	·184	·190	·196
63	·203	·209	·215	·222	·228	·235	·241	·248	·254	·261	·268	·274
64	·281	·288	·295	·302	·309	·316	·323	·330	·337	·344	·352	·359
65	2·366	2·374	2·381	2·389	2·396	2·404	2·411	2·419	2·427	2·435	2·443	2·451
66	·459	·467	·475	·483	·491	·499	·508	·516	·525	·533	·542	·551
67	·559	·568	·577	·586	·595	·604	·613	·622	·632	·641	·650	·660
68	·669	·679	·689	·699	·709	·718	·729	·739	·749	·759	·769	·780
69	·790	·801	·812	·823	·833	·844	·855	2·867	2·878	2·889	2·901	2·912
70	2·924	2·936	2·947	2·959	2·971	2·983	2·996	3·008	3·021	3·033	3·046	3·059

OF MINUTES AND SECONDS TO DECIMALS OF A DEGREE

″	0′	1′	2′	3′	4′	5′
0	0·0000	0·0167	0·0333	0·0500	0·0667	0·0833
1	003	169	336	503	669	836
2	006	172	339	506	672	839
3	008	175	342	508	675	842
4	011	178	344	511	678	844
5	0·0014	0·0181	0·0347	0·0514	0·0681	0·0847
6	017	183	350	517	683	850
7	019	186	353	519	686	853
8	022	189	356	522	689	856
9	025	192	358	525	692	858
10	0·0028	0·0194	0·0361	0·0528	0·0694	0·0861
11	031	197	364	531	697	864
12	033	200	367	533	700	867
13	036	203	369	536	703	869
14	039	206	372	539	706	872
15	0·0042	0·0208	0·0375	0·0542	0·0708	0·0875
16	044	211	378	544	711	878
17	047	214	381	547	714	881
18	050	217	383	550	717	883
19	053	219	386	553	719	886
20	0·0056	0·0222	0·0389	0·0556	0·0722	0·0889
21	058	225	392	558	725	892
22	061	228	394	561	728	894
23	064	231	397	564	731	897
24	067	233	400	567	733	900
25	0·0069	0·0236	0·0403	0·0569	0·0736	0·0903
26	072	239	406	572	739	906
27	075	242	408	575	742	908
28	078	244	411	578	744	911
29	081	247	414	581	747	914
30	0·0083	0·0250	0·0417	0·0583	0·0750	0·0917
31	086	253	419	586	753	919
32	089	256	422	589	756	922
33	092	258	425	592	758	925
34	094	261	428	594	761	928
35	0·0097	0·0264	0·0431	0·0597	0·0764	0·0931
36	100	267	433	600	767	933
37	103	269	436	603	769	936
38	106	272	439	606	772	939
39	108	275	442	608	775	942
40	0·0111	0·0278	0·0444	0·0611	0·0778	0·0944
41	114	281	447	614	781	947
42	117	283	450	617	783	950
43	119	286	453	619	786	953
44	122	289	456	622	789	956
45	0·0125	0·0292	0·0458	0·0625	0·0792	0·0958
46	128	294	461	628	794	961
47	131	297	464	631	797	964
48	133	300	467	633	800	967
49	136	303	469	636	803	969
50	0·0139	0·0306	0·0472	0·0639	0·0806	0·0972
51	142	308	475	642	808	975
52	144	311	478	644	811	978
53	147	314	481	647	814	981
54	150	317	483	650	817	983
55	0·0153	0·0319	0·0486	0·0653	0·0819	0·0986
56	156	322	489	656	822	989
57	158	325	492	658	825	992
58	161	328	494	661	828	994
59	0·0164	0·0331	0·0497	0·0664	0·0831	0·0997

Side proportional parts:

′	°
0	0·0
6	·1
12	·2
18	·3
24	·4
30	·5
36	·6
42	·7
48	·8
54	0·9

In units of the fourth decimal of a degree.

″	°
	0·0
0	
	·1
1	
	·5
2	
0·8	
	3
1·0	

In critical cases ascend.

OF HOURS, MINUTES AND SECONDS TO DECIMALS OF A DAY

m	0h	1h	2h	3h	4h	5h	s (SECONDS)	d
0	0·00000	0·04167	0·08333	0·12500	0·16667	0·20833	0	0·00000
1	·00069	·04236	·08403	·12569	·16736	·20903	1	·00001
2	·00139	·04306	·08472	·12639	·16806	·20972	2	·00002
3	·00208	·04375	·08542	·12708	·16875	·21042	3	·00003
4	·00278	·04444	·08611	·12778	·16944	·21111	4	·00005
5	0·00347	0·04514	0·08681	0·12847	0·17014	0·21181	5	0·00006
6	·00417	·04583	·08750	·12917	·17083	·21250	6	·00007
7	·00486	·04653	·08819	·12986	·17153	·21319	7	·00008
8	·00556	·04722	·08889	·13056	·17222	·21389	8	·00009
9	·00625	·04792	·08958	·13125	·17292·	·21458	9	·00010
10	0·00694	0·04861	0·09028	0·13194	0·17361	0·21528	10	0·00012
11	·00764	·04931	·09097	·13264	·17431	·21597	11	·00013
12	·00833	·05000	·09167	·13333	·17500	·21667	12	·00014
13	·00903	·05069	·09236	·13403	·17569	·21736	13	·00015
14	·00972	·05139	·09306	·13472	·17639	·21806	14	·00016
15	0·01042	0·05208	0·09375	0·13542	0·17708	0·21875	15	0·00017
16	·01111	·05278	·09444	·13611	·17778	·21944	16	·00019
17	·01181	·05347	·09514	·13681	·17847	·22014	17	·00020
18	·01250	·05417	·09583	·13750	·17917	·22083	18	·00021
19	·01319	·05486	·09653	·13819	·17986	·22153	19	·00022
20	0·01389	0·05556	0·09722	0·13889	0·18056	0·22222	20	0·00023
21	·01458	·05625	·09792	·13958	·18125	·22292	21	·00024
22	·01528	·05694	·09861	·14028	·18194	·22361	22	·00025
23	·01597	·05764	·09931	·14097	·18264	·22431	23	·00027
24	·01667	·05833	·10000	·14167	·18333	·22500	24	·00028
25	0·01736	0·05903	0·10069	0·14236	0·18403	0·22569	25	0·00029
26	·01806	·05972	·10139	·14306	·18472	·22639	26	·00030
27	·01875	·06042	·10208	·14375	·18542	·22708	27	·00031
28	·01944	·06111	·10278	·14444	·18611	·22778	28	·00032
29	·02014	·06181	·10347	·14514	·18681	·22847	29	·00034
30	0·02083	0·06250	0·10417	0·14583	0·18750	0·22917	30	0·00035
31	·02153	·06319	·10486	·14653	·18819	·22986	31	·00036
32	·02222	·06389	·10556	·14722	·18889	·23056	32	·00037
33	·02292	·06458	·10625	·14792	·18958	·23125	33	·00038
34	·02361	·06528	·10694	·14861	·19028	·23194	34	·00039
35	0·02431	0·06597	0·10764	0·14931	0·19097	0·23264	35	0·00041
36	·02500	·06667	·10833	·15000	·19167	·23333	36	·00042
37	·02569	·06736	·10903	·15069	·19236	·23403	37	·00043
38	·02639	·06806	·10972	·15139	·19306	·23472	38	·00044
39	·02708	·06875	·11042	·15208	·19375	·23542	39	·00045
40	0·02778	0·06944	0·11111	0·15278	0·19444	0·23611	40	0·00046
41	·02847	·07014	·11181	·15347	·19514	·23681	41	·00047
42	·02917	·07083	·11250	·15417	·19583	·23750	42	·00049
43	·02986	·07153	·11319	·15486	·19653	·23819	43	·00050
44	·03056	·07222	·11389	·15556	·19722	·23889	44	·00051
45	0·03125	0·07292	0·11458	0·15625	0·19792	0·23958	45	0·00052
46	·03194	·07361	·11528	·15694	·19861	·24028	46	·00053
47	·03264	·07431	·11597	·15764	·19931	·24097	47	·00054
48	·03333	·07500	·11667	·15833	·20000	·24167	48	·00056
49	·03403	·07569	·11736	·15903	·20069	·24236	49	·00057
50	0·03472	0·07639	0·11806	0·15972	0·20139	0·24306	50	0·00058
51	·03542	·07708	·11875	·16042	·20208	·24375	51	·00059
52	·03611	·07778	·11944	·16111	·20278	·24444	52	·00060
53	·03681	·07847	·12014	·16181	·20347	·24514	53	·00061
54	·03750	·07917	·12083	·16250	·20417	·24583	54	·00062
55	0·03819	0·07986	0·12153	0·16319	0·20486	0·24653	55	0·00064
56	·03889	·08056	·12222	·16389	·20556	·24722	56	·00065
57	·03958	·08125	·12292	·16458	·20625	·24792	57	·00066
58	·04028	·08194	·12361	·16528	·20694	·24861	58	·00067
59	0·04097	0·08264	0·12431	0·16597	0·20764	0·24931	59	·00068

For larger intervals of time, note that $6^h = 0^d \cdot 25$; $12^h = 0^d \cdot 50$; $18^h = 0^d \cdot 75$.

TABLE FOR CIRCUM-MERIDIAN OBSERVATIONS

CORRECTING FACTOR $m = 2 \sin^2 \tfrac{1}{2}(\text{H.A.}) \operatorname{cosec} 1''$

H.A. (m s)	m ($''$)	H.A. (m s)	m ($''$)	H.A. (m s)	m ($''$)	H.A. (m s)	m ($''$)	H.A. (m s)	m ($''$)	H.A. (m s)	m ($''$)
0 00	0	5 01·2	50	7 07·1	100	8 43·5	150	10 00	196	15 00	442
0 30	1	5 04·2	51	7 09·2	101	8 45·3	151	05	200	05	447
0 52	2	5 07·2	52	7 11·4	102	8 47·0	152	10	203	10	451
1 07	3	5 10·2	53	7 13·5	103	8 48·8	153	15	206	15	456
1 20	4	5 13·1	54	7 15·6	104	8 50·5	154	20	210	20	461
1 30	5	5 16·1	55	7 17·7	105	8 52·2	155	25	213	25	466
1 40	6	5 18·9	56	7 19·8	106	8 53·9	156	10 30	216	15 30	472
1 49	7	5 21·8	57	7 21·9	107	8 55·6	157	35	220	35	477
1 57	8	5 24·6	58	7 23·9	108	8 57·4	158	40	223	40	482
2 04	9	5 27·5	59	7 26·0	109	8 59·1	159	45	227	45	487
2 11	10	5 30·2	60	7 28·0	110	9 00·8	160	50	230	50	492
2 18	11	5 33·0	61	7 30·1	111	9 02·5	161	55	234	55	497
2 25	12	5 35·8	62	7 32·1	112	9 04·1	162	11 00	238	16 00	502
2 31	13	5 38·5	63	7 34·1	113	9 05·8	163	05	241	05	508
2 37	14	5 41·2	64	7 36·1	114	9 07·5	164	10	245	10	513
2 43	15	5 43·8	65	7 38·2	115	9 09·2	165	15	248	15	518
2 48	16	5 46·5	66	7 40·2	116	9 10·8	166	20	252	20	524
2 53	17	5 49·1	67	7 42·1	117	9 12·5	167	25	256	25	529
2 59	18	5 51·8	68	7 44·1	118	9 14·2	168	11 30	260	16 30	534
3 04	19	5 54·4	69	7 46·1	119	9 15·8	169	35	263	35	540
3 09	20	5 56·9	70	7 48·1	120	9 17·5	170	40	267	40	545
3 13	21	5 59·5	71	7 50·0	121	9 19·1	171	45	271	45	551
3 18	22	6 02·0	72	7 52·0	122	9 20·7	172	50	275	50	556
3 22	23	6 04·6	73	7 53·9	123	9 22·4	173	55	279	55	562
3 27	24	6 07·1	74	7 55·8	124	9 24·0	174	12 00	283	17 00	567
3 31	25	6 09·5	75	7 57·8	125	9 25·6	175	05	287	05	573
3 35	26	6 12·0	76	7 59·7	126	9 27·2	176	10	291	10	578
3 40	27	6 14·5	77	8 01·6	127	9 28·9	177	15	295	15	584
3 44	28	6 16·9	78	8 03·5	128	9 30·5	178	20	299	20	590
3 48	29	6 19·3	79	8 05·4	129	9 32·1	179	25	303	25	595
3 52	30	6 21·7	80	8 07·3	130	9 33·7	180	12 30	307	17 30	601
3 56	31	6 24·1	81	8 09·1	131	9 35·3	181	35	311	35	607
4 00	32	6 26·5	82	8 11·0	132	9 36·9	182	40	315	40	613
4 04	33	6 28·9	83	8 12·9	133	9 38·4	183	45	319	45	618
4 07	34	6 31·2	84	8 14·7	134	9 40·0	184	50	323	50	624
4 11	35	6 33·6	85	8 16·6	135	9 41·6	185	55	328	55	630
4 15	36	6 35·9	86	8 18·4	136	9 43·2	186	13 00	332	18 00	636
4 18	37	6 38·2	87	8 20·2	137	9 44·8	187	05	336	05	642
4 22	38	6 40·5	88	8 22·1	138	9 46·3	188	10	340	10	648
4 25	39	6 42·8	89	8 23·9	139	9 47·9	189	15	345	15	654
4 29	40	6 45·1	90	8 25·7	140	9 49·4	190	20	349	20	660
4 32	41	6 47·3	91	8 27·5	141	9 51·0	191	25	353	25	666
4 35	42	6 49·6	92	8 29·3	142	9 52·5	192	13 30	358	18 30	672
4 39	43	6 51·8	93	8 31·1	143	9 54·1	193	35	362	35	678
4 42	44	6 54·0	94	8 32·9	144	9 55·6	194	40	367	40	684
4 45	45	6 56·2	95	8 34·7	145	9 57·2	195	45	371	45	690
4 48	46	6 58·4	96	8 36·5	146	9 58·7	196	50	376	50	696
4 51	47	7 00·6	97	8 38·2	147	10 00·2	197	55	380	55	702
4 55	48	7 02·8	98	8 40·0	148	10 01·8	198	14 00	385	19 00	708
4 58	49	7 04·9	99	8 41·8	149	10 03·3	199	05	389	05	715
5 01		7 07·1		8 43·5		10 04·8		10	394	10	721
								15	399	15	727
								20	403	20	733
								25	408	25	740
								14 30	413	19 30	746
								35	417	35	753
								40	422	40	759
								45	427	45	765
								50	432	50	772
								55	437	55	778
								15 00	442	20 00	785

In critical cases ascend.

For further explanation see page xii.

m	0^h	1^h	2^h	3^h	4^h	5^h	s	$^s0$	$^s2$	$^s4$	$^s6$	$^s8$
	° ′	° ′	° ′	° ′	° ′	° ′		′ ″	′ ″	′ ″	′ ″	′ ″
0	0 00	15 00	30 00	45 00	60 00	75 00	0	0 00	0 03	0 06	0 09	0 12
1	0 15	15 15	30 15	45 15	60 15	75 15	1	0 15	0 18	0 21	0 24	0 27
2	0 30	15 30	30 30	45 30	60 30	75 30	2	0 30	0 33	0 36	0 39	0 42
3	0 45	15 45	30 45	45 45	60 45	75 45	3	0 45	0 48	0 51	0 54	0 57
4	1 00	16 00	31 00	46 00	61 00	76 00	4	1 00	1 03	1 06	1 09	1 12
5	1 15	16 15	31 15	46 15	61 15	76 15	5	1 15	1 18	1 21	1 24	1 27
6	1 30	16 30	31 30	46 30	61 30	76 30	6	1 30	1 33	1 36	1 39	1 42
7	1 45	16 45	31 45	46 45	61 45	76 45	7	1 45	1 48	1 51	1 54	1 57
8	2 00	17 00	32 00	47 00	62 00	77 00	8	2 00	2 03	2 06	2 09	2 12
9	2 15	17 15	32 15	47 15	62 15	77 15	9	2 15	2 18	2 21	2 24	2 27
10	2 30	17 30	32 30	47 30	62 30	77 30	10	2 30	2 33	2 36	2 39	2 42
11	2 45	17 45	32 45	47 45	62 45	77 45	11	2 45	2 48	2 51	2 54	2 57
12	3 00	18 00	33 00	48 00	63 00	78 00	12	3 00	3 03	3 06	3 09	3 12
13	3 15	18 15	33 15	48 15	63 15	78 15	13	3 15	3 18	3 21	3 24	3 27
14	3 30	18 30	33 30	48 30	63 30	78 30	14	3 30	3 33	3 36	3 39	3 42
15	3 45	18 45	33 45	48 45	63 45	78 45	15	3 45	3 48	3 51	3 54	3 57
16	4 00	19 00	34 00	49 00	64 00	79 00	16	4 00	4 03	4 06	4 09	4 12
17	4 15	19 15	34 15	49 15	64 15	79 15	17	4 15	4 18	4 21	4 24	4 27
18	4 30	19 30	34 30	49 30	64 30	79 30	18	4 30	4 33	4 36	4 39	4 42
19	4 45	19 45	34 45	49 45	64 45	79 45	19	4 45	4 48	4 51	4 54	4 57
20	5 00	20 00	35 00	50 00	65 00	80 00	20	5 00	5 03	5 06	5 09	5 12
21	5 15	20 15	35 15	50 15	65 15	80 15	21	5 15	5 18	5 21	5 24	5 27
22	5 30	20 30	35 30	50 30	65 30	80 30	22	5 30	5 33	5 36	5 39	5 42
23	5 45	20 45	35 45	50 45	65 45	80 45	23	5 45	5 48	5 51	5 54	5 57
24	6 00	21 00	36 00	51 00	66 00	81 00	24	6 00	6 03	6 06	6 09	6 12
25	6 15	21 15	36 15	51 15	66 15	81 15	25	6 15	6 18	6 21	6 24	6 27
26	6 30	21 30	36 30	51 30	66 30	81 30	26	6 30	6 33	6 36	6 39	6 42
27	6 45	21 45	36 45	51 45	66 45	81 45	27	6 45	6 48	6 51	6 54	6 57
28	7 00	22 00	37 00	52 00	67 00	82 00	28	7 00	7 03	7 06	7 09	7 12
29	7 15	22 15	37 15	52 15	67 15	82 15	29	7 15	7 18	7 21	7 24	7 27
30	7 30	22 30	37 30	52 30	67 30	82 30	30	7 30	7 33	7 36	7 39	7 42
31	7 45	22 45	37 45	52 45	67 45	82 45	31	7 45	7 48	7 51	7 54	7 57
32	8 00	23 00	38 00	53 00	68 00	83 00	32	8 00	8 03	8 06	8 09	8 12
33	8 15	23 15	38 15	53 15	68 15	83 15	33	8 15	8 18	8 21	8 24	8 27
34	8 30	23 30	38 30	53 30	68 30	83 30	34	8 30	8 33	8 36	8 39	8 42
35	8 45	23 45	38 45	53 45	68 45	83 45	35	8 45	8 48	8 51	8 54	8 57
36	9 00	24 00	39 00	54 00	69 00	84 00	36	9 00	9 03	9 06	9 09	9 12
37	9 15	24 15	39 15	54 15	69 15	84 15	37	9 15	9 18	9 21	9 24	9 27
38	9 30	24 30	39 30	54 30	69 30	84 30	38	9 30	9 33	9 36	9 39	9 42
39	9 45	24 45	39 45	54 45	69 45	84 45	39	9 45	9 48	9 51	9 54	9 57
40	10 00	25 00	40 00	55 00	70 00	85 00	40	10 00	10 03	10 06	10 09	10 12
41	10 15	25 15	40 15	55 15	70 15	85 15	41	10 15	10 18	10 21	10 24	10 27
42	10 30	25 30	40 30	55 30	70 30	85 30	42	10 30	10 33	10 36	10 39	10 42
43	10 45	25 45	40 45	55 45	70 45	85 45	43	10 45	10 48	10 51	10 54	10 57
44	11 00	26 00	41 00	56 00	71 00	86 00	44	11 00	11 03	11 06	11 09	11 12
45	11 15	26 15	41 15	56 15	71 15	86 15	45	11 15	11 18	11 21	11 24	11 27
46	11 30	26 30	41 30	56 30	71 30	86 30	46	11 30	11 33	11 36	11 39	11 42
47	11 45	26 45	41 45	56 45	71 45	86 45	47	11 45	11 48	11 51	11 54	11 57
48	12 00	27 00	42 00	57 00	72 00	87 00	48	12 00	12 03	12 06	12 09	12 12
49	12 15	27 15	42 15	57 15	72 15	87 15	49	12 15	12 18	12 21	12 24	12 27
50	12 30	27 30	42 30	57 30	72 30	87 30	50	12 30	12 33	12 36	12 39	12 42
51	12 45	27 45	42 45	57 45	72 45	87 45	51	12 45	12 48	12 51	12 54	12 57
52	13 00	28 00	43 00	58 00	73 00	88 00	52	13 00	13 03	13 06	13 09	13 12
53	13 15	28 15	43 15	58 15	73 15	88 15	53	13 15	13 18	13 21	13 24	13 27
54	13 30	28 30	43 30	58 30	73 30	88 30	54	13 30	13 33	13 36	13 39	13 42
55	13 45	28 45	43 45	58 45	73 45	88 45	55	13 45	13 48	13 51	13 54	13 57
56	14 00	29 00	44 00	59 00	74 00	89 00	56	14 00	14 03	14 06	14 09	14 12
57	14 15	29 15	44 15	59 15	74 15	89 15	57	14 15	14 18	14 21	14 24	14 27
58	14 30	29 30	44 30	59 30	74 30	89 30	58	14 30	14 33	14 36	14 39	14 42
59	14 45	29 45	44 45	59 45	74 45	89 45	59	14 45	14 48	14 51	14 54	14 57

The left-hand portion of the table gives the arc equivalent of each minute of time less than 6^h; for larger intervals note that $6^h = 90°$; $12^h = 180°$; $18^h = 270°$. The right-hand portion gives the equivalent of seconds, and fifths of seconds, of time.

INTERPOLATION TABLE FOR R

MUTUAL CONVERSION OF INTERVALS OF SOLAR AND SIDEREAL TIME

solar 0ʰ	ΔR	sidereal 0ʰ	solar 0ʰ	ΔR	sidereal 0ʰ	solar 1ʰ	ΔR	sidereal 1ʰ	solar 1ʰ	ΔR	sidereal 1ʰ
m s	s	m s	m s	s	m s	m s	s	m s	m s	s	m s
00 00.0	0.0	00 00.0	30 07.9	5.0	30 12.8	00 34.1	10.0	00 44.1	30 23.8	14.9	30 38.6
00 18.2	0.1	00 18.3	30 44.4	5.1	30 49.5	01 10.6	10.1	01 20.7	31 00.3	15.0	31 15.3
00 54.7	0.2	00 54.9	31 20.9	5.2	31 26.1	01 47.2	10.2	01 57.3	31 36.8	15.1	31 51.9
01 31.3	0.3	01 31.5	31 57.5	5.3	32 02.7	02 23.7	10.3	02 33.9	32 13.4	15.2	32 28.5
02 07.8	0.4	02 08.1	32 34.0	5.4	32 39.3	03 00.2	10.4	03 10.6	32 49.9	15.3	33 05.1
02 44.3	0.5	02 44.8	33 10.5	5.5	33 16.0	03 36.7	10.5	03 47.2	33 26.4	15.4	33 41.8
03 20.8	0.6	03 21.4	33 47.0	5.6	33 52.6	04 13.3	10.6	04 23.8	34 02.9	15.5	34 18.4
03 57.4	0.7	03 58.0	34 23.6	5.7	34 29.2	04 49.8	10.7	05 00.4	34 39.5	15.6	34 55.0
04 33.9	0.8	04 34.6	35 00.1	5.8	35 05.8	05 26.3	10.8	05 37.1	35 16.0	15.7	35 31.6
05 10.4	0.9	05 11.3	35 36.6	5.9	35 42.5	06 02.8	10.9	06 13.7	35 52.5	15.8	36 08.3
05 46.9	1.0	05 47.9	36 13.1	6.0	36 19.1	06 39.4	11.0	06 50.3	36 29.0	15.9	36 44.9
06 23.5	1.1	06 24.5	36 49.7	6.1	36 55.7	07 15.9	11.1	07 26.9	37 05.6	16.0	37 21.5
07 00.0	1.2	07 01.1	37 26.2	6.2	37 32.3	07 52.4	11.2	08 03.6	37 42.1	16.1	37 58.1
07 36.5	1.3	07 37.8	38 02.7	6.3	38 09.0	08 28.9	11.3	08 40.2	38 18.6	16.2	38 34.8
08 13.0	1.4	08 14.4	38 39.2	6.4	38 45.6	09 05.4	11.4	09 16.8	38 55.1	16.3	39 11.4
08 49.6	1.5	08 51.0	39 15.8	6.5	39 22.2	09 42.0	11.5	09 53.4	39 31.7	16.4	39 48.0
09 26.1	1.6	09 27.6	39 52.3	6.6	39 58.8	10 18.5	11.6	10 30.0	40 08.2	16.5	40 24.6
10 02.6	1.7	10 04.2	40 28.8	6.7	40 35.5	10 55.0	11.7	11 06.7	40 44.7	16.6	41 01.3
10 39.1	1.8	10 40.9	41 05.3	6.8	41 12.1	11 31.5	11.8	11 43.3	41 21.2	16.7	41 37.9
11 15.6	1.9	11 17.5	41 41.9	6.9	41 48.7	12 08.1	11.9	12 19.9	41 57.8	16.8	42 14.5
11 52.2	2.0	11 54.1	42 18.4	7.0	42 25.3	12 44.6	12.0	12 56.5	42 34.3	16.9	42 51.1
12 28.7	2.1	12 30.7	42 54.9	7.1	43 02.0	13 21.1	12.1	13 33.2	43 10.8	17.0	43 27.8
13 05.2	2.2	13 07.4	43 31.4	7.2	43 38.6	13 57.6	12.2	14 09.8	43 47.3	17.1	44 04.4
13 41.7	2.3	13 44.0	44 08.0	7.3	44 15.2	14 34.2	12.3	14 46.4	44 23.9	17.2	44 41.0
14 18.3	2.4	14 20.6	44 44.5	7.4	44 51.8	15 10.7	12.4	15 23.0	45 00.4	17.3	45 17.6
14 54.8	2.5	14 57.2	45 21.0	7.5	45 28.5	15 47.2	12.5	15 59.7	45 36.9	17.4	45 54.3
15 31.3	2.6	15 33.9	45 57.5	7.6	46 05.1	16 23.7	12.6	16 36.3	46 13.4	17.5	46 30.9
16 07.8	2.7	16 10.5	46 34.1	7.7	46 41.7	17 00.3	12.7	17 12.9	46 50.0	17.6	47 07.5
16 44.4	2.8	16 47.1	47 10.6	7.8	47 18.3	17 36.8	12.8	17 49.5	47 26.5	17.7	47 44.1
17 20.9	2.9	17 23.7	47 47.1	7.9	47 55.0	18 13.3	12.9	18 26.2	48 03.0	17.8	48 20.7
17 57.4	3.0	18 00.4	48 23.6	8.0	48 31.6	18 49.8	13.0	19 02.8	48 39.5	17.9	48 57.4
18 33.9	3.1	18 37.0	49 00.1	8.1	49 08.2	19 26.4	13.1	19 39.4	49 16.0	18.0	49 34.0
19 10.5	3.2	19 13.6	49 36.7	8.2	49 44.8	20 02.9	13.2	20 16.0	49 52.6	18.1	50 10.6
19 47.0	3.3	19 50.2	50 13.2	8.3	50 21.4	20 39.4	13.3	20 52.7	50 29.1	18.2	50 47.2
20 23.5	3.4	20 26.9	50 49.7	8.4	50 58.1	21 15.9	13.4	21 29.3	51 05.6	18.3	51 23.9
21 00.0	3.5	21 03.5	51 26.2	8.5	51 34.7	21 52.5	13.5	22 05.9	51 42.1	18.4	52 00.5
21 36.6	3.6	21 40.1	52 02.8	8.6	52 11.3	22 29.0	13.6	22 42.5	52 18.7	18.5	52 37.1
22 13.1	3.7	22 16.7	52 39.3	8.7	52 47.9	23 05.5	13.7	23 19.2	52 55.2	18.6	53 13.7
22 49.6	3.8	22 53.4	53 15.8	8.8	53 24.6	23 42.0	13.8	23 55.8	53 31.7	18.7	53 50.4
23 26.1	3.9	23 30.0	53 52.3	8.9	54 01.2	24 18.6	13.9	24 32.4	54 08.2	18.8	54 27.0
24 02.7	4.0	24 06.6	54 28.9	9.0	54 37.8	24 55.1	14.0	25 09.0	54 44.8	18.9	55 03.6
24 39.2	4.1	24 43.2	55 05.4	9.1	55 14.4	25 31.6	14.1	25 45.7	55 21.3	19.0	55 40.2
25 15.7	4.2	25 19.9	55 41.9	9.2	55 51.1	26 08.1	14.2	26 22.3	55 57.8	19.1	56 16.9
25 52.2	4.3	25 56.5	56 18.4	9.3	56 27.7	26 44.7	14.3	26 58.9	56 34.3	19.2	56 53.5
26 28.8	4.4	26 33.1	56 55.0	9.4	57 04.3	27 21.2	14.4	27 35.5	57 10.9	19.3	57 30.1
27 05.3	4.5	27 09.7	57 31.5	9.5	57 40.9	27 57.7	14.5	28 12.1	57 47.4	19.4	58 06.7
27 41.8	4.6	27 46.4	58 08.0	9.6	58 17.6	28 34.2	14.6	28 48.8	58 23.9	19.5	58 43.4
28 18.3	4.7	28 23.0	58 44.5	9.7	58 54.2	29 10.7	14.7	29 25.4	59 00.4	19.6	59 20.0
28 54.9	4.8	28 59.6	59 21.1	9.8	59 30.8	29 47.3	14.8	30 02.0	59 37.0	19.7	59 56.6
29 31.4	4.9	29 36.2	59 57.6	9.9	60 07.4	30 23.8		30 38.6	60 13.5		60 33.2
30 07.9		30 12.8	60 34.1		60 44.1						

In critical cases ascend.

Add ΔR to solar time interval (left-hand argument) to obtain sidereal time interval.
Subtract ΔR from sidereal time interval (right-hand argument) to obtain solar time interval.

MUTUAL CONVERSION OF INTERVALS OF SOLAR AND SIDEREAL TIME

solar 2ʰ	ΔR	sidereal 2ʰ	solar 2ʰ	ΔR	sidereal 2ʰ	solar 3ʰ	ΔR	sidereal 3ʰ	solar 3ʰ	ΔR	sidereal 3ʰ
m s	s	m s	m s	s	m s	m s	s	m s	m s	s	m s
00 13·5	19·8	00 33·2	30 03·2	24·7	30 27·8	00 29·4	29·7	00 59·0	30 19·1	34·6	30 53·6
00 50·0	19·9	01 09·9	30 39·7	24·8	31 04·4	01 05·9	29·8	01 35·7	30 55·6	34·7	31 30·2
01 26·5	20·0	01 46·5	31 16·2	24·9	31 41·1	01 42·4	29·9	02 12·3	31 32·1	34·8	32 06·9
02 03·1	20·1	02 23·1	31 52·7	25·0	32 17·7	02 19·0	30·0	02 48·9	32 08·6	34·9	32 43·5
02 39·6	20·2	02 59·7	32 29·3	25·1	32 54·3	02 55·5	30·1	03 25·5	32 45·2	35·0	33 20·1
03 16·1	20·3	03 36·4	33 05·8	25·2	33 30·9	03 32·0	30·2	04 02·2	33 21·7	35·1	33 56·7
03 52·6	20·4	04 13·0	33 42·3	25·3	34 07·6	04 08·5	30·3	04 38·8	33 58·2	35·2	34 33·4
04 29·2	20·5	04 49·6	34 18·8	25·4	34 44·2	04 45·1	30·4	05 15·4	34 34·7	35·3	35 10·0
05 05·7	20·6	05 26·2	34 55·4	25·5	35 20·8	05 21·6	30·5	05 52·0	35 11·3	35·4	35 46·6
05 42·2	20·7	06 02·9	35 31·9	25·6	35 57·4	05 58·1	30·6	06 28·6	35 47·8	35·5	36 23·2
06 18·7	20·8	06 39·5	36 08·4	25·7	36 34·1	06 34·6	30·7	07 05·3	36 24·3	35·6	36 59·9
06 55·2	20·9	07 16·1	36 44·9	25·8	37 10·7	07 11·1	30·8	07 41·9	37 00·8	35·7	37 36·5
07 31·8	21·0	07 52·7	37 21·5	25·9	37 47·3	07 47·7	30·9	08 18·5	37 37·4	35·8	38 13·1
08 08·3	21·1	08 29·3	37 58·0	26·0	38 23·9	08 24·2	31·0	08 55·1	38 13·9	35·9	38 49·7
08 44·8	21·2	09 06·0	38 34·5	26·1	39 00·6	09 00·7	31·1	09 31·8	38 50·4	36·0	39 26·4
09 21·3	21·3	09 42·6	39 11·0	26·2	39 37·2	09 37·2	31·2	10 08·4	39 26·9	36·1	40 03·0
09 57·9	21·4	10 19·2	39 47·6	26·3	40 13·8	10 13·8	31·3	10 45·0	40 03·5	36·2	40 39·6
10 34·4	21·5	10 55·8	40 24·1	26·4	40 50·4	10 50·3	31·4	11 21·6	40 40·0	36·3	41 16·2
11 10·9	21·6	11 32·5	41 00·6	26·5	41 27·1	11 26·8	31·5	11 58·3	41 16·5	36·4	41 52·9
11 47·4	21·7	12 09·1	41 37·1	26·6	42 03·7	12 03·3	31·6	12 34·9	41 53·0	36·5	42 29·5
12 24·0	21·8	12 45·7	42 13·7	26·7	42 40·3	12 39·9	31·7	13 11·5	42 29·6	36·6	43 06·1
13 00·5	21·9	13 22·3	42 50·2	26·8	43 16·9	13 16·4	31·8	13 48·1	43 06·1	36·7	43 42·7
13 37·0	22·0	13 59·0	43 26·7	26·9	43 53·6	13 52·9	31·9	14 24·8	43 42·6	36·8	44 19·4
14 13·5	22·1	14 35·6	44 03·2	27·0	44 30·2	14 29·4	32·0	15 01·4	44 19·1	36·9	44 56·0
14 50·1	22·2	15 12·2	44 39·8	27·1	45 06·8	15 06·0	32·1	15 38·0	44 55·6	37·0	45 32·6
15 26·6	22·3	15 48·8	45 16·3	27·2	45 43·4	15 42·5	32·2	16 14·6	45 32·2	37·1	46 09·2
16 03·1	22·4	16 25·5	45 52·8	27·3	46 20·0	16 19·0	32·3	16 51·3	46 08·7	37·2	46 45·8
16 39·6	22·5	17 02·1	46 29·3	27·4	46 56·7	16 55·5	32·4	17 27·9	46 45·2	37·3	47 22·5
17 16·2	22·6	17 38·7	47 05·8	27·5	47 33·3	17 32·1	32·5	18 04·5	47 21·7	37·4	47 59·1
17 52·7	22·7	18 15·3	47 42·4	27·6	48 09·9	18 08·6	32·6	18 41·1	47 58·3	37·5	48 35·7
18 29·2	22·8	18 52·0	48 18·9	27·7	48 46·5	18 45·1	32·7	19 17·8	48 34·8	37·6	49 12·3
19 05·7	22·9	19 28·6	48 55·4	27·8	49 23·2	19 21·6	32·8	19 54·4	49 11·3	37·7	49 49·0
19 42·3	23·0	20 05·2	49 31·9	27·9	49 59·8	19 58·2	32·9	20 31·0	49 47·8	37·8	50 25·6
20 18·8	23·1	20 41·8	50 08·5	28·0	50 36·4	20 34·7	33·0	21 07·6	50 24·4	37·9	51 02·2
20 55·3	23·2	21 18·5	50 45·0	28·1	51 13·0	21 11·2	33·1	21 44·3	51 00·9	38·0	51 38·8
21 31·8	23·3	21 55·1	51 21·5	28·2	51 49·7	21 47·7	33·2	22 20·9	51 37·4	38·1	52 15·5
22 08·4	23·4	22 31·7	51 58·0	28·3	52 26·3	22 24·3	33·3	22 57·5	52 13·9	38·2	52 52·1
22 44·9	23·5	23 08·3	52 34·6	28·4	53 02·9	23 00·8	33·4	23 34·1	52 50·5	38·3	53 28·7
23 21·4	23·6	23 45·0	53 11·1	28·5	53 39·5	23 37·3	33·5	24 10·8	53 27·0	38·4	54 05·3
23 57·9	23·7	24 21·6	53 47·6	28·6	54 16·2	24 13·8	33·6	24 47·4	54 03·5	38·5	54 42·0
24 34·5	23·8	24 58·2	54 24·1	28·7	54 52·8	24 50·3	33·7	25 24·0	54 40·0	38·6	55 18·6
25 11·0	23·9	25 34·8	55 00·7	28·8	55 29·4	25 26·9	33·8	26 00·6	55 16·6	38·7	55 55·2
25 47·5	24·0	26 11·5	55 37·2	28·9	56 06·0	26 03·4	33·9	26 37·2	55 53·1	38·8	56 31·8
26 24·0	24·1	26 48·1	56 13·7	29·0	56 42·7	26 39·9	34·0	27 13·9	56 29·6	38·9	57 08·5
27 00·5	24·2	27 24·7	56 50·2	29·1	57 19·3	27 16·4	34·1	27 50·5	57 06·1	39·0	57 45·1
27 37·1	24·3	28 01·3	57 26·8	29·2	57 55·9	27 53·0	34·2	28 27·1	57 42·7	39·1	58 21·7
28 13·6	24·4	28 37·9	58 03·3	29·3	58 32·5	28 29·5	34·3	29 03·7	58 19·2	39·2	58 58·3
28 50·1	24·5	29 14·6	58 39·8	29·4	59 09·2	29 06·0	34·4	29 40·4	58 55·7	39·3	59 35·0
29 26·6	24·6	29 51·2	59 16·3	29·5	59 45·8	29 42·5	34·5	30 17·0	59 32·2	39·4	60 11·6
30 03·2		30 27·8	59 52·9	29·6	60 22·4	30 19·1		30 53·6	60 08·8		60 48·2
			60 29·4		60 59·0						

In critical cases ascend.

Add ΔR to solar time interval (left-hand argument) to obtain sidereal time interval.
Subtract ΔR from sidereal time interval (right-hand argument) to obtain solar time interval.

INTERPOLATION TABLE FOR R

MUTUAL CONVERSION OF INTERVALS OF SOLAR AND SIDEREAL TIME

solar 4^h	ΔR	sidereal 4^h
m s	s	m s
00 08.8	39.5	00 48.2
00 45.3	39.6	01 24.8
01 21.8	39.7	02 01.5
01 58.3	39.8	02 38.1
02 34.9	39.9	03 14.7
03 11.4	40.0	03 51.3
03 47.9	40.1	04 28.0
04 24.4	40.2	05 04.6
05 00.9	40.3	05 41.2
05 37.5	40.4	06 17.8
06 14.0	40.5	06 54.4
06 50.5	40.6	07 31.1
07 27.0	40.7	08 07.7
08 03.6	40.8	08 44.3
08 40.1	40.9	09 20.9
09 16.6	41.0	09 57.6
09 53.1	41.1	10 34.2
10 29.7	41.2	11 10.8
11 06.2	41.3	11 47.4
11 42.7	41.4	12 24.1
12 19.2	41.5	13 00.7
12 55.8	41.6	13 37.3
13 32.3	41.7	14 13.9
14 08.8	41.8	14 50.6
14 45.3	41.9	15 27.2
15 21.9	42.0	16 03.8
15 58.4	42.1	16 40.4
16 34.9	42.2	17 17.1
17 11.4	42.3	17 53.7
17 48.0	42.4	18 30.3
18 24.5	42.5	19 06.9
19 01.0	42.6	19 43.6
19 37.5	42.7	20 20.2
20 14.1	42.8	20 56.8
20 50.6	42.9	21 33.4
21 27.1	43.0	22 10.1
22 03.6	43.1	22 46.7
22 40.2	43.2	23 23.3
23 16.7	43.3	23 59.9
23 53.2	43.4	24 36.5
24 29.7	43.5	25 13.2
25 06.2	43.6	25 49.8
25 42.8	43.7	26 26.4
26 19.3	43.8	27 03.0
26 55.8	43.9	27 39.7
27 32.3	44.0	28 16.3
28 08.9	44.1	28 52.9
28 45.4	44.2	29 29.5
29 21.9	44.3	30 06.2
29 58.4	44.4	30 42.8
30 35.0		31 19.4

solar 4^h	ΔR	sidereal 4^h
m s	s	m s
30 35.0	44.5	31 19.4
31 11.5	44.6	31 56.0
31 48.0	44.7	32 32.7
32 24.5	44.8	33 09.3
33 01.1	44.9	33 45.9
33 37.6	45.0	34 22.5
34 14.1	45.1	34 59.2
34 50.6	45.2	35 35.8
35 27.2	45.3	36 12.4
36 03.7	45.4	36 49.0
36 40.2	45.5	37 25.7
37 16.7	45.6	38 02.3
37 53.3	45.7	38 38.9
38 29.8	45.8	39 15.5
39 06.3	45.9	39 52.2
39 42.8	46.0	40 28.8
40 19.4	46.1	41 05.4
40 55.9	46.2	41 42.0
41 32.4	46.3	42 18.7
42 08.9	46.4	42 55.3
42 45.4	46.5	43 31.9
43 22.0	46.6	44 08.5
43 58.5	46.7	44 45.1
44 35.0	46.8	45 21.8
45 11.5	46.9	45 58.4
45 48.1	47.0	46 35.0
46 24.6	47.1	47 11.6
47 01.1	47.2	47 48.3
47 37.6	47.3	48 24.9
48 14.2	47.4	49 01.5
48 50.7	47.5	49 38.1
49 27.2	47.6	50 14.8
50 03.7	47.7	50 51.4
50 40.3	47.8	51 28.0
51 16.8	47.9	52 04.6
51 53.3	48.0	52 41.3
52 29.8	48.1	53 17.9
53 06.4	48.2	53 54.5
53 42.9	48.3	54 31.1
54 19.4	48.4	55 07.8
54 55.9	48.5	55 44.4
55 32.5	48.6	56 21.0
56 09.0	48.7	56 57.6
56 45.5	48.8	57 34.3
57 22.0	48.9	58 10.9
57 58.6	49.0	58 47.5
58 35.1	49.1	59 24.1
59 11.6	49.2	60 00.8
59 48.1	49.3	60 37.4
60 24.7		61 14.0

solar 5^h	ΔR	sidereal 5^h
m s	s	m s
00 24.7	49.4	01 14.0
01 01.2	49.5	01 50.6
01 37.7	49.6	02 27.3
02 14.2	49.7	03 03.9
02 50.7	49.8	03 40.5
03 27.3	49.9	04 17.1
04 03.8	50.0	04 53.7
04 40.3	50.1	05 30.4
05 16.8	50.2	06 07.0
05 53.4	50.3	06 43.6
06 29.9	50.4	07 20.2
07 06.4	50.5	07 56.9
07 42.9	50.6	08 33.5
08 19.5	50.7	09 10.1
08 56.0	50.8	09 46.7
09 32.5	50.9	10 23.4
10 09.0	51.0	11 00.0
10 45.6	51.1	11 36.6
11 22.1	51.2	12 13.2
11 58.6	51.3	12 49.9
12 35.1	51.4	13 26.5
13 11.7	51.5	14 03.1
13 48.2	51.6	14 39.7
14 24.7	51.7	15 16.4
15 01.2	51.8	15 53.0
15 37.8	51.9	16 29.6
16 14.3	52.0	17 06.2
16 50.8	52.1	17 42.9
17 27.3	52.2	18 19.5
18 03.9	52.3	18 56.1
18 40.4	52.4	19 32.7
19 16.9	52.5	20 09.4
19 53.4	52.6	20 46.0
20 30.0	52.7	21 22.6
21 06.5	52.8	21 59.2
21 43.0	52.9	22 35.9
22 19.5	53.0	23 12.5
22 56.0	53.1	23 49.1
23 32.6	53.2	24 25.7
24 09.1	53.3	25 02.3
24 45.6	53.4	25 39.0
25 22.1	53.5	26 15.6
25 58.7	53.6	26 52.2
26 35.2	53.7	27 28.8
27 11.7	53.8	28 05.5
27 48.2	53.9	28 42.1
28 24.8	54.0	29 18.7
29 01.3	54.1	29 55.3
29 37.8	54.2	30 32.0
30 14.3		31 08.6

solar 5^h	ΔR	sidereal 5^h
m s	s	m s
30 14.3	54.3	31 08.6
30 50.9	54.4	31 45.2
31 27.4	54.5	32 21.8
32 03.9	54.6	32 58.5
32 40.4	54.7	33 35.1
33 17.0	54.8	34 11.7
33 53.5	54.9	34 48.3
34 30.0	55.0	35 25.0
35 06.5	55.1	36 01.6
35 43.1	55.2	36 38.2
36 19.6	55.3	37 14.8
36 56.1	55.4	37 51.5
37 32.6	55.5	38 28.1
38 09.2	55.6	39 04.7
38 45.7	55.7	39 41.3
39 22.2	55.8	40 18.0
39 58.7	55.9	40 54.6
40 35.3	56.0	41 31.2
41 11.8	56.1	42 07.8
41 48.3	56.2	42 44.4
42 24.8	56.3	43 21.1
43 01.3	56.4	43 57.7
43 37.9	56.5	44 34.3
44 14.4	56.6	45 10.9
44 50.9	56.7	45 47.6
45 27.4	56.8	46 24.2
46 04.0	56.9	47 00.8
46 40.5	57.0	47 37.4
47 17.0	57.1	48 14.1
47 53.5	57.2	48 50.7
48 30.1	57.3	49 27.3
49 06.6	57.4	50 03.9
49 43.1	57.5	50 40.6
50 19.6	57.6	51 17.2
50 56.2	57.7	51 53.8
51 32.7	57.8	52 30.4
52 09.2	57.9	53 07.1
52 45.7	58.0	53 43.7
53 22.3	58.1	54 20.3
53 58.8	58.2	54 56.9
54 35.3	58.3	55 33.6
55 11.8	58.4	56 10.2
55 48.4	58.5	56 46.8
56 24.9	58.6	57 23.4
57 01.4	58.7	58 00.1
57 37.9	58.8	58 36.7
58 14.5	58.9	59 13.3
58 51.0	59.0	59 49.9
59 27.5	59.1	60 26.6
60 04.0		61 03.2

In critical cases ascend.

Add ΔR to solar time interval (left-hand argument) to obtain sidereal time interval.
Subtract ΔR from sidereal time interval (right-hand argument) to obtain solar time interval.

Time diff.	Tabular six-hourly difference																							
	3	6	9	12	15	18	21	24	27	30	33	36	39	42	45	48	51	54	57	60	63	66	69	72
h m																								
0 10	0	0	0	0	0	1	1	1	1	1	1	1	1	1	1	1	1	2	2	2	2	2	2	2
20	0	0	1	1	1	1	1	1	2	2	2	2	2	2	3	3	3	3	3	3	4	4	4	4
30	0	1	1	1	1	2	2	2	2	3	3	3	3	4	4	4	4	5	5	5	5	6	6	6
40	0	1	1	1	2	2	2	3	3	3	4	4	4	5	5	5	6	6	6	7	7	7	8	8
50	0	1	1	2	2	3	3	3	4	4	5	5	5	6	6	7	7	8	8	8	9	9	10	10
1 00	1	1	2	2	3	3	4	4	5	5	6	6	7	7	8	8	9	9	10	10	11	11	12	12
10	1	1	2	2	3	4	4	5	5	6	6	7	8	8	9	9	10	11	11	12	12	13	13	14
20	1	1	2	3	3	4	5	5	6	7	7	8	9	9	10	11	11	12	13	13	14	15	15	16
30	1	2	2	3	4	5	5	6	7	8	8	9	10	11	11	12	13	14	14	15	16	17	17	18
40	1	2	3	3	4	5	6	7	8	8	9	10	11	12	13	13	14	15	16	17	18	18	19	20
50	1	2	3	4	5	6	6	7	8	9	10	11	12	13	14	15	16	17	17	18	19	20	21	22
2 00	1	2	3	4	5	6	7	8	9	10	11	12	13	14	15	16	17	18	19	20	21	22	23	24
10	1	2	3	4	5	7	8	9	10	11	12	13	14	15	16	17	18	20	21	22	23	24	25	26
20	1	2	4	5	6	7	8	9	11	12	13	14	15	16	18	19	20	21	22	23	25	26	27	28
30	1	3	4	5	6	8	9	10	11	13	14	15	16	18	19	20	21	23	24	25	26	28	29	30
40	1	3	4	5	7	8	9	11	12	13	15	16	17	19	20	21	23	24	25	27	28	29	31	32
50	1	3	4	6	7	9	10	11	13	14	16	17	18	20	21	23	24	26	27	28	30	31	33	34
3 00	2	3	5	6	8	9	11	12	14	15	17	18	20	21	23	24	26	27	29	30	32	33	35	36
10	2	3	5	6	8	10	11	13	14	16	17	19	21	22	24	25	27	29	30	32	33	35	36	38
20	2	3	5	7	8	10	12	13	15	17	18	20	22	23	25	27	28	30	32	33	35	37	38	40
30	2	4	5	7	9	11	12	14	16	18	19	21	23	25	26	28	30	32	33	35	37	39	40	42
40	2	4	6	7	9	11	13	15	17	18	20	22	24	26	28	29	31	33	35	37	39	40	42	44
50	2	4	6	8	10	12	13	15	17	19	21	23	25	27	29	31	33	35	36	38	40	42	44	46
4 00	2	4	6	8	10	12	14	16	18	20	22	24	26	28	30	32	34	36	38	40	42	44	46	48
10	2	4	6	8	10	13	15	17	19	21	23	25	27	29	31	33	35	38	40	42	44	46	48	50
20	2	4	7	9	11	13	15	17	20	22	24	26	28	30	33	35	37	39	41	43	46	48	50	52
30	2	5	7	9	11	14	16	18	20	23	25	27	29	32	34	36	38	41	43	45	47	50	52	54
40	2	5	7	9	12	14	16	19	21	23	26	28	30	33	35	37	40	42	44	47	49	51	54	56
50	2	5	7	10	12	15	17	19	22	24	27	29	31	34	36	39	41	44	46	48	51	53	56	58
5 00	3	5	8	10	13	15	18	20	23	25	28	30	33	35	38	40	43	45	48	50	53	55	58	60
10	3	5	8	10	13	16	18	21	23	26	28	31	34	36	39	41	44	47	49	52	54	57	59	62
20	3	5	8	11	13	16	19	21	24	27	29	32	35	37	40	43	45	48	51	53	56	59	61	64
30	3	6	8	11	14	17	19	22	25	28	30	33	36	39	41	44	47	50	52	55	58	61	63	66
40	3	6	9	11	14	17	20	23	26	28	31	34	37	40	43	45	48	51	54	57	60	62	65	68
50	3	6	9	12	15	18	20	23	26	29	32	35	38	41	44	47	50	53	55	58	61	64	67	70
6 00	3	6	9	12	15	18	21	24	27	30	33	36	39	42	45	48	51	54	57	60	63	66	69	72

INTERPOLATION TABLE FOR STARS

Day of month	Monthly difference																				
	1	2	3	4	5	6	7	8	9	10	11	12	14	16	18	20	22	24	26	28	30
2	0	0	0	0	0	0	0	1	1	1	1	1	1	1	1	1	1	2	2	2	2
4	0	0	0	1	1	1	1	1	1	1	1	2	2	2	2	3	3	3	3	4	4
6	0	0	1	1	1	1	1	2	2	2	2	2	3	3	4	4	4	5	5	6	6
8	0	1	1	1	1	2	2	2	2	3	3	3	4	4	5	5	6	6	7	7	8
10	0	1	1	1	2	2	2	3	3	3	4	4	5	5	6	7	7	8	9	9	10
12	0	1	1	2	2	2	3	3	4	4	4	5	6	6	7	8	9	10	10	11	12
14	0	1	1	2	2	3	3	4	4	5	5	6	7	7	8	9	10	11	12	13	14
16	1	1	2	2	3	3	4	4	5	5	6	6	7	9	10	11	12	13	14	15	16
18	1	1	2	2	3	3	4	5	5	6	7	7	8	10	11	12	13	14	16	17	18
20	1	1	2	3	3	4	5	5	6	7	7	8	9	11	12	13	15	16	17	19	20
22	1	1	2	3	4	4	5	6	7	7	8	9	10	12	13	15	16	18	19	21	22
24	1	2	2	3	4	5	6	6	7	8	9	10	11	13	14	16	18	19	21	22	24
26	1	2	3	3	4	5	6	7	8	9	10	10	12	14	16	17	19	21	23	24	26
28	1	2	3	4	5	6	7	7	8	9	10	11	13	15	17	19	21	22	24	26	28
30	1	2	3	4	5	6	7	8	9	10	11	12	14	16	18	20	22	24	26	28	30

As an alternative to the main tabulations of R and Sun on pages 2–25, monthly sets of polynomial coefficients are given on the opposite page. The coefficients have been included to enable users who possess small programmable electronic calculators to evaluate R and Sun directly without recourse to interpolation tables.

Formulae for the evaluation of the polynomials are given beneath the tabulations. The polynomial series for Dec and E may be evaluated efficiently by using the second expression in which the interpolating factor x is used as a constant multiplier and the formation of the separate powers of x is avoided; the algorithm may be written as

$$b_{n+1} = b_n x + a_{4-n}, \text{ for } n = 1 \text{ to } 4$$

where $b_1 = a_4$ and $b_5 = $ required value. Particular care must be taken to ensure that the coefficients are entered with the correct signs; check sums are provided for checking that the coefficients have been entered correctly.

Example for 2002 January 6^d 13^h 23^m $49^s\!.3$ UT.

$x = 6.558\ 2095/32 = 0.204\ 9440$

$R = +6^h\!.632\ 410 + 2^h\!.102\ 731\ x = 7^h\!.063\ 352 = 7^h\ 03^m\ 48^s$

	Dec				E		
b_1	$-$	$0.114\ 04$	b_2 $-0.267\ 6918$	b_1	$-$	$0.016\ 068$	b_2 $+0.057\ 9450$
b_3	$+$	$3.909\ 9782$	b_4 $+3.092\ 5167$	b_3	$+$	$0.042\ 1665$	b_4 $-0.244\ 7912$
b_5	$-$	$22.474\ 3771$	$= S\ 22°\ 28'\ 28''$	b_5	$+$	$11.902\ 8745$	$= 11^h\ 54^m\ 10^s$

$SD = +0^s\!.271\ 61 - 0^s\!.000\ 55\ x = 0^s\!.271\ 497 = 16'\ 17''$

Parallax. If the full accuracy of the polynomial coefficients is to be used then it is necessary to correct the observed altitude of the Sun for parallax by adding $8''\!.8 \cos$ (altitude).

Accuracy of the Polynomial Method. Estimates of the standard error and the maximum error that can arise using the polynomials are given below.

	R	Dec	E	SD
Standard error	$0^s\!.012$	$0''\!.19$	$0^s\!.031$	$0''\!.15$
Maximum error	$0^s\!.065$	$0''\!.50$	$0^s\!.075$	$0''\!.35$

The error distribution is approximately Gaussian over a long period of time, but errors on a given day for a particular quantity are likely to be systematic. If the values from the polynomials are rounded before use to $1''$ for Dec and SD and $0^s\!.1$ for R and E, then additional rounding errors are introduced, and the maximum possible errors are increased to $1''\!.0$ for Dec, $0''\!.9$ for SD, and $0^s\!.12$ for R and $0^s\!.13$ for E.

Conversion of GST to UT. For planning purposes it may be useful to find the approximate values of UT corresponding to given values of GST on a particular day d. The series for R may be used. First calculate GST at 0^h UT on the day concerned from $a_0 + (d/32)a_1$.

Then $\qquad$ UT $= (GST - GST \text{ at } 0^h \text{ UT}) \times 0.997\ 2696$

since a mean value of the rate of increase of GST with UT may be used.

For example, suppose we wish to find the UT corresponding to $00^h\ 24^m\ 08^s$ GST on 2002 January 6.

$\qquad$ GST at 0^h UT on 2002 January 6 $= 7^h\!.026\ 672$

(or the value $7^h\ 01^m\ 36^s\!.0$ may be taken directly from the main tabulation on page 2). Hence

$\qquad$ UT $= (24^h\!.402\ 222 - 7^h\!.026\ 672) \times 0.997\ 2696 = 17^h\!.328\ 108 = 17^h\ 19^m\ 41^s$

where 24^h has been added to GST to make the term in parentheses positive.

		R		R		R		R
		h		h		h		h
a_0	Jan.	+ 6·632410	Apr.	+12·546280	July	+18·525895	Oct.	+ 0·571189
a_1		+ 2·102731		+ 2·102702		+ 2·102730		+ 2·102701
a_0	Feb.	+ 8·669432	May	+14·517563	Aug.	+20·562914	Nov.	+ 2·608180
a_1		+ 2·102704		+ 2·102724		+ 2·102709		+ 2·102723
a_0	Mar.	+10·509300	June	+16·554578	Sept.	+22·599913	Dec.	+ 4·579484
a_1		+ 2·102690		+ 2·102737		+ 2·102695		+ 2·102742

		Dec	E		Dec	E		Dec	E
		°	h		°	h		°	h
a_0	Jan.	−23·10817	+11·953043	May	+14·64958	+12·045563	Sept.	+ 8·78257	+11·991640
a_1		+ 2·29119	− 0·253433		+ 9·85568	+ 0·072762		−11·49907	+ 0·164586
a_2		+ 3·96484	+ 0·030291		− 2·01325	− 0·081704		− 1·23269	+ 0·042753
a_3		− 0·24432	+ 0·061238		− 0·56696	− 0·003481		+ 0·55390	− 0·018510
a_4		− 0·11404	− 0·016068		+ 0·07379	+ 0·005514		− 0·02572	− 0·006215
Check sum		−17·21050	+11·775071		+21·99884	+12·038654		− 3·42101	+12·174254
a_0	Feb.	−17·49017	+11·777472	June	+21·85745	+12·041045	Oct.	− 2·64474	+12·163460
a_1		+ 8·86888	− 0·080677		+ 4·62761	− 0·074519		−12·43360	+ 0·175347
a_2		+ 2·68371	+ 0·110393		− 3·22004	− 0·070156		+ 0·19066	− 0·038927
a_3		− 0·66851	+ 0·006862		− 0·33127	+ 0·042184		+ 0·60111	− 0·026092
a_4		+ 0·01476	− 0·011426		+ 0·13196	− 0·003332		− 0·01409	− 0·000645
Check sum		− 6·59133	+11·802624		+23·06571	+11·935222		−14·30066	+12·273143
a_0	Mar.	− 8·11441	+11·789295	July	+23·19466	+11·941699	Nov.	−13·97678	+12·272513
a_1		+12·06478	+ 0·098209		− 1·84535	− 0·105215		−10·42066	+ 0·024042
a_2		+ 1·03941	+ 0·070610		− 3·48996	+ 0·023228		+ 1·87331	− 0·112021
a_3		− 0·62734	− 0·022743		+ 0·18263	+ 0·049736		+ 0·69644	− 0·012587
a_4		+ 0·02387	− 0·002734		+ 0·07279	− 0·015170		− 0·06571	+ 0·008587
Check sum		+ 4·38631	+11·932637		+18·11477	+11·894278		−21·89340	+12·180534
a_0	Apr.	+ 3·99954	+11·927607	Aug.	+18·36182	+11·893425	Dec.	−21·57660	+12·192848
a_1		+12·39787	+ 0·161283		− 7·82695	+ 0·024862		− 5·28910	− 0·190706
a_2		− 0·58651	− 0·017270		− 2·60480	+ 0·081286		+ 3·49804	− 0·100405
a_3		− 0·58333	− 0·023191		+ 0·50569	+ 0·009630		+ 0·48758	+ 0·044588
a_4		+ 0·02997	+ 0·001359		− 0·01371	− 0·012391		− 0·17244	+ 0·000758
Check sum		+15·25754	+12·049788		+ 8·42205	+11·996812		−23·05252	+11·947083

SUN'S SEMI-DIAMETER (SD)

		°		°		°		°
a_0	Jan.	+0·27161	Apr.	+0·26727	July	+0·26255	Oct.	+0·26658
a_1		−0·00055		−0·00235		+0·00043		+0·00241
a_0	Feb.	+0·27107	May	+0·26503	Aug.	+0·26296	Nov.	+0·26896
a_1		−0·00173		−0·00176		+0·00153		+0·00190
a_0	Mar.	+0·26957	June	+0·26328	Sept.	+0·26444	Dec.	+0·27078
a_1		−0·00234		−0·00073		+0·00226		+0·00083

The interpolating factor $x = d/32$, where d is the sum of the day of the month and the decimal of the day in UT.

R and SD are computed from $a_0 + a_1 x$ where a_0, a_1 are given in the table.

Dec and E are computed from the polynomial:

$$a_0 + a_1 x + a_2 x^2 + a_3 x^3 + a_4 x^4 \equiv (((a_4 x + a_3)x + a_2)x + a_1)x + a_0$$

The coefficients for each month are valid for $0 \le x \le 1$. To obtain full accuracy x should be calculated to seven decimal places.

NORTHERN STARS

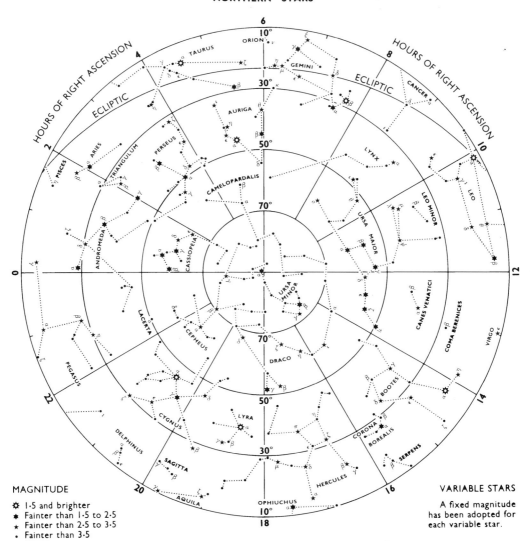

EQUATORIAL STARS (R.A. 12ʰ to 24ʰ)

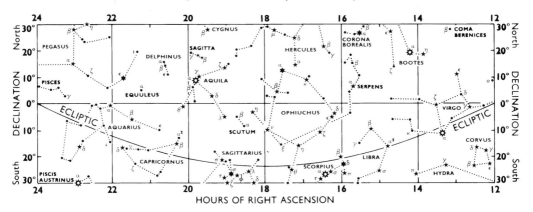

HOURS OF RIGHT ASCENSION

SOUTHERN STARS

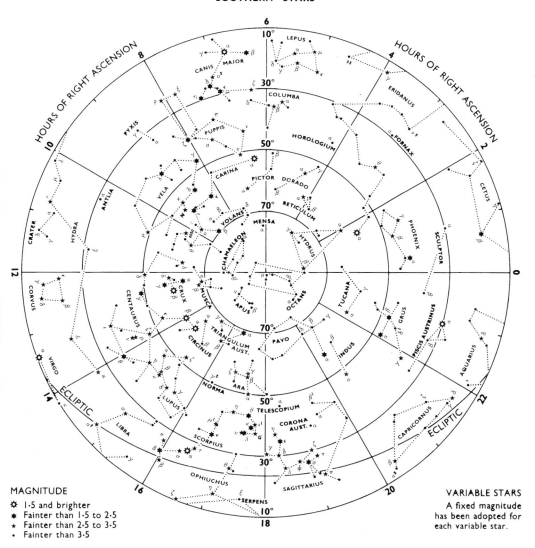

MAGNITUDE

✦ 1·5 and brighter
★ Fainter than 1·5 to 2·5
⋆ Fainter than 2·5 to 3·5
• Fainter than 3·5

VARIABLE STARS

A fixed magnitude
has been adopted for
each variable star.

EQUATORIAL STARS (R.A. 0ʰ to 12ʰ)

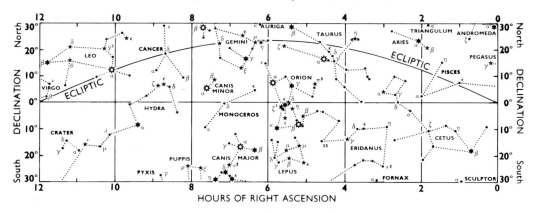

HOURS OF RIGHT ASCENSION

This index gives the number by which each star is distinguished in the list of apparent places on pages 26-53.

The five northern circumpolar stars are indicated by NP, the five southern by SP. Apparent places of these ten stars (for every ten days) are tabulated on pages 54-55.

The numbers of some of the brighter stars that are known by their proper names are as follows:

Proper Name	Number	Proper Name	Number	Proper Name	Number
Achernar	32	*Canopus*	179	*Pollux*	216
Aldebaran	116	*Capella*	136	*Procyon*	212
Algol	70	*Castor*	210	*Regulus*	273
Altair	548	*Deneb*	571	*Rigel*	135
Antares	441	*Denebola*	316	*Sirius*	185
Arcturus	369	*Dubhe*	298	*Spica*	353
Bellatrix	140	*Fomalhaut*	632	*Vega*	514
Betelgeuse	162	*Polaris*	NP		

Name No.	Name No.	Name No.	Name No.	Name No.	Name No.
Andromedae	**Aquarii**	**Aurigae**	**Cancri**	**Capricorni**	**Cassiopeiae**
α 1	θ 611	α 136	α 244	ι 588	50 44
β 24	ι 605	β 166	β 224	ψ 573	
γ 45	λ 630	δ 165	δ 237	ω 577	**Centauri**
δ 12	τ 627	ε 127	ι 239		α 379
ζ 15	φ 639	ζ 128		**Carinae**	β 364
ι 646	88 637	η 131	**Canis Majoris**	α 179	γ 337
κ 649	98 642	θ 167	α 185	β 253	δ 320
λ 645		ι 126	β 177	ε 228	ε 356
μ 19	**Aquilae**	κ 172	γ 195	θ 291	ζ 362
ν 17	α 548	ν 159	δ 196	ι 255	η 378
ο 634	β 551		ε 191	υ 268	θ 367
υ 30	γ 545	**Bootis**	ζ 174	χ 220	ι 351
51 31	δ 540	α 369	η 205	ω 275	κ 391
	ε 526	β 392	θ 190	I 282	λ 311
Antliae	ζ 529	γ 377	κ 187	a 252	μ 360
α 284	η 549	δ 396	ο² 193	l 267	ν 359
	θ 558	ε 385	σ 192	u 295	ξ² 347
Apodis	λ 530	ζ 380	ω 200	BS 3571 243	π 307
α 386		η 361		BS 4050 279	σ 329
β 683	**Arae**	θ 374	**Canis Minoris**	BS 4114 286	ψ 373
γ 443	α 479	λ 371	α 212	BS 4140 287	d 354
δ¹ 682	β 470	μ 404	β 207	BS 4337 299	υ 372
η 679	γ 471	ρ 376			I 357
R 681	δ 475		**Canum Venat.**	**Cassiopeiae**	BS 4522 315
	ε¹ 461	**Camelopardalis**	α 344	α 13	BS 4889 341
Aquarii	ζ 460	α 124	β 334	β 2	BS 5485 384
α 604	η 452	β 129		γ 18	
β 593	θ 499	BS 1035 78	**Capricorni**	δ 26	**Cephei**
γ 613		BS 1686 656	α² 561	ε 37	α 587
δ 631	**Arietis**	BS 2527 657	β 562	ζ 11	β 592
ε 576	α 46	BS 3082 658	γ 595	η 16	γ 647
ζ 614	β 39	BS 4646 661	δ 601	κ 10	δ 616
η 619	41 61	BS 4893 662	ζ 591	23 651	ζ 610
			θ 581	49 652	η 574

Name	No.	Name	No.	Name	No.	Name	No.	Name	No.	Name	No.
Muscae		**Pavonis**		**Pictoris**		**Sagittarii**		**Tauri**		**Ursae Majoris**	
λ	312	α	564	α	186	θ^1	555	γ	110	χ	314
Normae		β	572	β	157	ι	550	δ	111	ψ	300
γ^2	436	γ	590	γ	158	λ	512	ϵ	113	23	260
		δ	556			μ	502	ζ	150	**Ursae Minoris**	
Octantis		ϵ	554	**Piscis Aust.**		ξ^2	524	η	90	α	NP
α	684	ζ	515	α	632	o	528	θ^2	114	β	389
β	624	η	487	β	617	π	533	λ	99	γ	402
δ	680	λ	521	ϵ	620	ρ	537	μ	107	δ	NP
ϵ	685	ξ	507	ι	599	σ	522	ν	100	ϵ	453
ζ	SP					τ	531	ξ	77	ζ	417
θ	670	**Pegasi**		**Piscium**		ϕ	516	o	76	η	664
ι	SP	α	636	α	43			τ	120	4	663
ν	596	β	635	γ	640	**Scorpii**		5	79	5	375
σ	SP	γ	4	ϵ	21	α	441	10	82	**Velorum**	
τ	SP	ϵ	598	η	28	β	430	17	86	γ	223
χ	SP	ζ	621	θ	643	δ	428	27	92	δ	238
		η	623	ι	648	ϵ	454			κ	258
Ophiuchi		θ	609	ω	650	ζ^2	457	**Telescopii**		λ	251
α	481	ι	606			η	465	α	511	μ	292
β	486	λ	625	**Puppis**		θ	482			o	234
γ	490	μ	629	ζ	221	ι^1	489	**Trianguli**		ϕ	271
δ	433	π	608	ν	182	κ	485	α	36	ψ	261
ϵ	435	I	589	ξ	218	λ	480	β	47	N	262
ζ	448			π	201	μ^1	455	γ	49	BS 3426	233
η	464	**Persei**		ρ	222	μ^2	456			BS 3445	235
θ	469	α	75	σ	209	π	426	**Triang. Aust.**		BS 3591	247
ι	458	β	70	τ	188	σ	438	α	451	BS 3614	250
κ	459	γ	68	I	199	τ	447	β	423	BS 4023	276
λ	444	δ	83	c	217	υ	476	γ	399	BS 4180	290
ν	496	ϵ	95	BS 3080	219	G	491	δ	434		
σ	473	ζ	94	BS 3270	225	N	445	ϵ	412	**Virginis**	
44	472	η	62							α	353
45	474	θ	58	**Pyxidis**		**Sculptoris**		**Tucanae**		β	317
67	497	ι	71	α	236	α	20	α	612	γ	338
72	500	λ	101	γ	241	β	644	γ	641	δ	343
		μ	104	**Reticuli**				ζ	6	ϵ	346
Orionis		ν	87	α	106	**Scuti**				ζ	355
α	162	ξ	97	β	88	α	513	**Ursae Majoris**		η	326
β	135	o	85	δ	98	β	518	α	298	θ	348
γ	140	ρ	69					β	297	ι	370
δ	144	τ	63	**Sagittae**		**Serpentis**		γ	318	κ	368
ϵ	149	ϕ	33	β	543	α	416	δ	323	μ	383
ζ^1	153	48	102	γ	553	β	418	ϵ	342	ν	313
η	139			δ	546	γ	424	ζ	352	o	319
ι	148					δ	409	η	358	τ	363
κ	156	**Phoenicis**		**Sagittarii**		ϵ	422	θ	263	109	387
λ	147	α	9	α	539	η	506	ι	245		
ν	170	β	22	β^1	538	κ	419	κ	248	**Volantis**	
π^3	122	γ	27	γ	498	μ	420	λ	278	α	249
π^4	123	δ	29	δ	505	ξ	483	μ	281	β	230
π^5	125	ϵ	3	ϵ	509			ν	304	γ^2	197
σ	151	κ	8	ζ	527	**Tauri**		ξ	303	δ	203
τ	137	ψ	38	η	503	α	116	o	231	ζ	214
						β	141	υ	269		

Printed in the United Kingdom by The Stationery Office Limited, London, England

TJ003842 C15 04/01 19585 610029